OUTLINES OF PHYSICAL GEOGRAPHY

OUTLINES OF PHYSICAL GEOGRAPHY

Mamta Sharma

ABHIJEET PUBLICATIONS

New Delhi-110002

ABHIJEET PUBLICATIONS
4658-A, 1st Floor, Ambika Bhawan
21 Ansari Road, Daryaganj
New Delhi-110002
Phone: 011-23259444, 65698474
E-Mail: abhijeetpublication@gmail.com

OUTLINES OF PHYSICAL GEOGRAPHY

First Published 2012

ISBN: 978-93-81136-90-4

PRINTED AT INDIA

Published by J.K. Singh for Abhijeet Publications, New Delhi-110002, Laser Typeset by Piyush Computers, New Delhi and Printed at Asian Offset, Delhi.

PREFACE

Over geological time the earth has changed physically. A product of these changes was the creation of conditions in which life became possible. Once established, the burgeoning life-forms which the earth supported became a part of the substantive matrix of earth evolution. Some of the latest-developed of these opportunistic life-forms were small, bipedal apes. From the forests and savannahs of Africa, some of the hominid ancestors of these apes evolved into people. As people, and like the other life-forms with which they shared the planet, they responded in different ways to the environment of which they had become a part. And out of these responses, empowered by the development of language and bonded by the crystallization of ritual into religion, their influence upon it began to enlarge. Human groups became culturally differentiated: some became sedentary as the realization of agricultural systems and incipient urbanization signalled the possibilities of societal enlargement and the extension of socio-political ambition through the control of other people and of earth space. It was some of these groups that were to be primarily responsible for 'The Great Transformation'. Physical Geography is a sub field of geography that focuses on the systematic study of patterns and processes within the hydrosphere biosphere atmosphere and lithosphere. It aims to understand the physical layout of the earth its weather and global flora and fauna patterns. Many areas of physical geography make use of geology particularly in the study of weathering and erosion. The geology of other planets is discussed at Geological features of the solar system. Physical geography as a scientific discipline is usually contrasted with and complemented by its sister science human geography.

'The Great Transformation' is a topical theme in geography.

The argument runs that the ecosphere has suffered changes of such magnitude and variety at the hands of the human species that its fundamental appearance and nature have been, or are about to be, radically altered. It would be wrong to belittle this view; indeed, much evidence in its favour will emerge later in the encyclopedia. None the less, it is right from the full perspective of geological time to note that the Great Transformation is but one of the changes, though possibly one of the most radical, that have occurred in the ecosphere during its history. All living beings, and not just the human species, actively change their environment. Since life first evolved, the living and non-living worlds have interacted. The course of biological evolution has not been steered solely by environmental change: life itself changes its environment and to a large extent determines the conditions in the air, sea, and land surface where it thrives.

In practice, and in relatively recent times, this view has been severely strained by the widening gap between physical and human geography, by the tensions created between competing explanatory paradigms—especially in human geography—and by a multiplicity of steadily narrowing specialisms. Superficially, the outcome may seem to threaten the academic integrity and coherence of the discipline. Yet this coherence is arguably more necessary now than at any time in the past, as habitat pressures strain the weakening seams of a stretched earth. This encyclopedia, therefore, offers a view of geography from an earlier, wider, and sometimes bolder perspective by bringing together a wide variety of contributions from scholars of both the physical and human environments, and by focusing their expertise on the large theme of the evolution of the earth as a habitat.

—Author

Contents

1

HISTORY OF PHYSICAL GEOGRAPHY

Geographic knowledge saw strong growth in Europe and the United States in the 1800s. This period also saw the emergence of a number of societies interested in geographic issues. In Germany, Alexander von Humboldt, Carl Ritter, and Fredrich Ratzel made substantial contributions to human and physical geography. Humboldt's publication Kosmos (1844) examines the geology and physical geography of the Earth. This work is considered by many academics to be a milestone contribution to geographic scholarship. Late in the 19th Century, Ratzel theorized that the distribution and culture of the Earth's various human populations was strongly influenced by the natural environment. The French geographer Paul Vidal de la Blanche opposed this revolutionary idea.

Instead, he suggested that human beings were a dominant force shaping the form of the environment. The idea that humans were modifying the physical environment was also prevalent in the United States. In 1847, George Perkins Marsh gave an address to the Agricultural Society of Rutland County, Vermont. The subject of this speech was that human activity was having a destructive impact on land, especially through deforestation and land conversion. This speech also became the foundation for his book Man and Nature or The Earth as Modified by Human Action, first published in 1864. In this publication, Marsh warned of the ecological consequences of the continued development of the American frontier.

During the first 50 years of the 1900s, many academics in the field of geography extended the various ideas presented in the previous century to studies of small regions all over the world. Most of these studies used descriptive field methods to test research questions. Starting in about 1950, geographic research experienced a shift in methodology. Geographers began adopting a more scientific approach that relied on quantitative techniques. The quantitative revolution was also associated with a change in the way in which geographers studied the Earth and its phenomena. Researchers now began investigating process rather than mere description of the event of interest. Today, the quantitative approach is becoming even more prevalent due to advances in computer and software technologies.

In 1964, William Pattison published an article in the Journal of Geography (1964, 63: 211-216) that suggested that modern Geography was now composed of the following four academic traditions:

- Spatial Tradition - the investigation of the phenomena of geography from a strictly spatial perspective.
- Area Studies Tradition - the geographical study of an area on the Earth at either the local, regional, or global scale.
- Human-Land Tradition - the geographical study of human interactions with the environment.
- Earth Science Tradition - the study of natural phenomena from a spatial perspective. This tradition is best described as theoretical physical geography.

Today, the academic traditions described by Pattison are still dominant fields of geographical investigation. However, the frequency and magnitude of human mediated environmental problems has been on a steady increase since the publication of this notion. These increases are the result of a growing human population and the consequent increase in the consumption of natural resources. As a result, an increasing number of researchers in geography are studying how humans modify the environment. A significant number of these projects also develop strategies to reduce the negative impact of human activities on nature. Some of the dominant themes in these studies include: environmental degradation of the hydrosphere, atmosphere, lithosphere, and biosphere; resource use issues; natural hazards; environmental impact assessment; and the effect of urbanization and land-use change on natural environments. Considering all of the statements presented concerning the history and development of geography, we are now ready to formulate a somewhat coherent definition. This definition suggests that geography, in its simplest form, is the field of knowledge that is concerned with how phenomena are spatially organized. Physical geography attempts to determine why natural phenomena have particular spatial patterns and orientation. This online textbook will focus primarily on the Earth Science Tradition. Some of the information that is covered in this textbook also deals with the alterations of the environment because of human interaction. These pieces of information belong in the Human-Land Tradition of geography.

History of Physical Geography

The nature of understanding in physical geography has changed over time. When investigating this change it becomes apparent that certain universal ideas or forces had very important ramifications to the academic study of physical geography. During the period from 1850 to 1950, there seems to be four main ideas that had a strong influence on the discipline:

1. *Uniformitarianism* - this theory rejected the idea that catastrophic forces were responsible for the current conditions on the Earth. It suggested instead that continuing uniformity of existing processes were responsible for the present and past conditions of this planet.
2. *Evolution* - Charles Darwin's Origin of Species (1859) suggested that natural selection determined which individuals would pass on their genetic traits to future generations. As a result of this theory, evolutionary explanations for a variety of natural phenomena were postulated by scientists. The theories of uniformitarianism and evolution arose from a fundamental change in the way humans explained the universe and nature. During the 16th, 17th, and 18th

centuries scholars began refuting belief or myth based explanations of the cosmos, and instead used science to help explain the mysteries of nature. Belief based explanations of the cosmos are made consistent with a larger framework of knowledge that focuses on some myth. However, theories based on science questioned the accuracy of these beliefs.

3. *Exploration and Survey* - much of the world had not been explored before 1900. Thus, during this period all of the fields of physical geography were actively involved with basic data collection. This data collection included activities like determining the elevation of land surfaces, classification and description of landforms, the measurement of the volume of flow of rivers, measurement of various phenomena associated to weather and climate, and the classification of soils, organisms, biological communities and ecosystems.
4. *Conservation* - beginning in the 1850s a concern for the environment began to develop as a result of the human development of once natural areas in the United States and Europe. One of the earliest statements of these ideas came from George Perkins Marsh (1864) in his book "Man in Nature" or "Physical Geography as Modified by Human Action". This book is often cited by scholars as the first significant academic contribution to conservation and environmentalism.

After 1950, the following two forces largely determined the nature of physical geography:

1. *The Quantitative Revolution* - measurement became the central focus of research in Physical Geography. It was used primarily for hypothesis testing. With measurement came mapping, models, statistics, mathematics, and hypothesis testing. The quantitative revolution was also associated with a change in the way in which physical geographers studied the Earth and its phenomena. Researchers now began investigating process rather than mere description of the environment.

2. *The study of Human/Land Relationships* - the influence of human activity on the environment was becoming very apparent after 1950. As a result, many researchers in physical geography began studying the influence of humans on the environment. Some of the dominant themes in these studies included: environmental degradation and resource use; natural hazards and impact assessment; and the effect of urbanization and land-use change on natural environments.

Another alternative set of theories concerning landscapes and landforms presumed that change could be understood as time-independent. In these theories, forms did not change over time, there was instead a steady balance between variables that produced forms that could not be placed in a temporal sequence.

ELEMENTS OF GEOGRAPHY

In the previous section, we discovered that geography consists of at least two different sub-fields of knowledge with similar methodology: Physical geography and human geography.

Table: Some of the phenomena studied in physical and human geography.

Physical Geography	Human Geography
Rocks and Minerals	Population
Landforms	Settlements
Soils	Economic Activities
Animals	Transportation
Plants	Recreational Activities
Water	Religion
Atmosphere	Political Systems
Rivers and Other Water Bodies	Social Traditions
Environment	Human Migration
Climate and Weather	Agricultural Systems
Oceans	Urban Systems

The following table also helps to make the differences between these two types of geography more apparent. This table describes some of the phenomena or elements studied by each of these sub-fields of knowledge. Knowing what kinds of things are studied by geographers provides us with a better understanding of the differences between physical and human geography.

The graphic model in above Figure indicates that the study of geography can also involve a holistic synthesis. Holistic synthesis connects knowledge from a variety of academic fields in both human and physical geography. For example, the study of the enhancement of the Earth's greenhouse effect and the resulting global warming requires a multidisciplinary approach for complete understanding. The fields of climatology and meteorology are required to understand the physical effects of adding addition greenhouse gases to the atmosphere's radiation balance. The field of economic geography provides information on how various forms of human economic activity contribute to the emission of greenhouse gases through fossil fuel burning and land-use change. Combining the knowledge of both of these academic areas gives us a more comprehensive understanding of why this serious environmental problem occurs.

The holistic nature of geography is both a strength and a weakness. Geography's strength comes from its ability to connect functional interrelationships that are not normally noticed in narrowly defined fields of knowledge. The most obvious weakness associated with the geographical approach is related to the fact that holistic understanding is often too simple and misses important details of cause and effect.

Scope of Physical Geography

We have now learned that physical geography examines and investigates natural phenomena spatially. In the previous section, we identified some of the key elements studied by physical geographers. Combining these two items, we can now suggest that physical geography studies the spatial patterns of weather and climate, soils, vegetation, animals, water in all its forms, and landforms. Physical geography also examines the interrelationships of these phenomena to human activities. This sub-field of geography is academically known as the Human-Land Tradition. This area of geography has seen very keen interest and growth in the last few decades because of the acceleration of human induced environmental degradation. Thus, physical geography's scope is much broader than the simple spatial study of nature. It also involves the investigation of how humans are influencing nature.

Academics studying physical geography and other related earth sciences are rarely generalists. Most are in fact highly specialized in their fields of knowledge and tend to focus themselves in one of the following well defined areas of understanding in physical geography:

- Geomorphology - studies the various landforms on the Earth's surface.
- Pedology - is concerned with the study of soils.
- Biogeography - is the science that investigates the spatial relationships of plants and animals.
- Hydrology - is interested in the study of water in all its forms.
- Meteorology - studies the circulation of the atmosphere over short time spans.
- Climatology - studies the effects of weather on life and examines the circulation of the atmosphere over longer time spans.

The above fields of knowledge generally have a primary role in introductory textbooks dealing with physical geography. Introductory physical geography textbooks can also contain information from other related disciplines including:

- Geology - studies the form of the Earth's surface and subsurface, and the processes that create and modify it.
- Ecology - the scientific study of the interactions between organisms and their environment.
- Oceanography - the science that examines the biology, chemistry, physics, and geology of oceans.

- Cartography - the technique of making maps.
- Astronomy - the science that examines celestial bodies and the cosmos.

Geography as an Environmental Science

Webster's 9th Collegiate Dictionary defines Environment "... as the complex of physical, chemical, and biotic factors (such as climate, soil, and living things) that act upon an organism or an ecological community and ultimately determines its form and survival".

Both human and physical geography provide an important intellectual background for studying the environment. Many environmental studies/science programs offered by Universities and Colleges around the world rely on the information found in various geography courses to help educate their students about the state of the environment.

Gilbert's concept of dynamic equilibrium (1877) and Hack's model of the same (1960) are classic examples of this approach. Both Gilbert and Hack viewed spatial variations as more important than temporal variations. Spatial variability replaced temporal variability as the dominant feature of these theories. In this view, the landscape is in balance with current inputs and so there is no temporal memory in the landscape.

The only possible sources of variation in the landscape are not the different stages in development or even past histories, but instead spatial variations in the nature of and inputs into the landscape. Although the exclusion of the 'past' as a source of variation in the landscape is a problem in time-independent concepts, they do form a basis for trying to understand the landscape as a product of contemporary processes. These models bring landscape-forming processes to the time- and space-scale of the investigators and detach the landscapes from their unknowable (in human terms) pasts. They provide, in other words, a platform for examining the landscape at spatial and temporal scales relevant for applications such as engineering.

The above two approaches to landscape study are very Anglo-American in their orientation. There were other theoretical frameworks within which landscapes could be

explained. Two of the most significant on continental Europe were climatic and structural geomorphology. Climatic geomorphology, as the name suggests, viewed climate as the overarching explanatory framework for landscapes. This is the view that climate controlled the characteristics and distribution of landforms. Climate is the controlling factor in landscape development. Climate resulted in a distinct latitudinal zonation of landforms, with an additional similar variation for the effect of altitude. There was a consistent causal relationship between the two, between climate and landscape, based on the persistent and consistent presence of a set of landforming processes unique to each climatic environment.

A zonation of landforms on the basis of climate was also developed by Tricart and Cailleux (1972) in France. The climatic zones, and indeed continents, were not static and this was reflected in the landforms found in different regions. Landforms and landscape could reflect the action of past climates, and so the simple latitudinal zonation had superimposed 'historical' patterns of climatic influence.

Linton's (1955) model of tor development on Dartmoor was a classic example of the application of climate and its change as the explanation for the development of a specific landform. Linton hypothesized a two-stage climatically controlled series of processes.

First, the core stones of the tors were developed through the deep weathering of granite under tropical conditions. Large core stones were produced at depth where weathering was less intense and smaller core stones near the surface where alteration was greatest.

The regolith remained around the core stones until removed by an episode of periglacial activity. This climate was more effective at removing weathered material, and with this removed the core stones formed disjointed masses on hilltops and valley sides. The idea of 'natural' zonation of the physical environment, particularly as controlled by climate, was applied not only to geomorphology. Herbertson (1905) developed his idea of natural regions based on the distribution of biota under the influence of climate on an ideal continent.

Peltier's (1950) version of the importance of climate related more to the relationship between process and climate than directly to landforms. Peltier insisted that climate controlled the ability of processes to operate.

A particular combination of temperature and precipitation would invariably result in the operation of mechanical or chemical processes of denudation or a particular combination of both. Although this vision of geomorphology was developed, and still influences geomorphology on continental Europe, it found little favour in the UK, as noted by Gregory (1985, quoting Stoddart, 1968; Derbyshire, 1973; Douglas, 1980).

The main criticism seems to have been that the idea of climate affecting landform and landscape development was perfectly acceptable, but the idea that it was the major theory for landscape development was not. The view of a specific climate resulting in a specific landscape was seen as too simplistic and not able to cope with the complexities of interpretation that structural differences and past histories imposed on it.

These early trends in the classification and quantification of natural phenomena tend not to be mentioned in 'standard' histories of physical geography. The reasons why are unclear. One possible reason is that these measurements were often made without the use of an overarching explanatory framework within which to place the measurements.

Gregory (1985), for example, notes that Kuchler categorized the early twentieth century as a period of data collection and accumulation during which there was no co-ordinating or overarching theory with which to explain the patterns observed. Within subdisciplines such as meteorology, however, this was not the case as theories of atmospheric circulation and weather patterns were available to co-ordinate and explain temperature patterns, windflows and other climatic parameters. The recent book by Solomon (2001), for example, illustrates that there were accepted explanatory frameworks based on detailed theories for weather patterns even in the data-poor Antarctic region at the end of the nineteenth century.

These theories were used to predict the likely behaviour of weather patterns and for planning expeditions, although the extreme and unpredicted weather of 1912 resulted in the tragedy of Scott's failed expedition. The early twentieth century would see these patterns and classifications extended by the likes of Koppen, and later by Thornwaite (1948), but the numerical basis for these schemes was developed in the nineteenth century.

By the 1960s the questions that physical geography was asking of the environment had altered, and the explanatory frameworks used had to evolve as well. Davis's cycle of erosion focused research onto the megascale, onto questions of landscape development over eons.

Pressing questions of engineering concern, such as how is a landslide going to behave if a road is cut through it, were not answerable within this theoretical framework. Whether this type of question was appropriate for this framework, and by implication an appropriate question for geomorphology, was not, it seems, a concern. The nature of questions in physical geography had altered since Davis posed questions on a 'grand' scale. The quest for relevance and the need to justify their academic existence meant a focusing of geomorphology on questions of relevance to society.

Whether these questions were questions of relevance to everyone or to particular groups in particular locations has not been made clear in the histories of the subject. Of central concern, however, was the recognition that most questions concerning landscape on a human spatial and temporal scale were just unanswerable as well as unaskable using the Davisian framework.

Identifying a planation surface did not help in identifying the stability of a specific hillslope, nor did the identification of a section of the landscape as mature help in identifying and predicting rates of change. The nature of questions of interest to geomorphologists had altered, and so their operating framework had to change.

Other parts of physical geography had long discussed issues at scales other than the regional and geologic. Climate studies, for example, had integrated short-term measurements of changes in variables such as temperature into explanatory frameworks at a range of scales. Prediction of weather patterns relied on the rapid collection and processing of information within changing models of how the atmosphere operated. Small-scale and large-scale phenomena were both recognized and modelled using the same physical laws. Aggregation of behaviour was needed to predict at the regional and global scale, but the principles on which the models were based were similar.

Ecological studies had likewise developed a scale-independent framework for analysing and integrating information. The ecosystem concept had been forcefully put forward by Tansley (1935). The ability to identify, classify, define and relate entities within the landscape at any scale made this form of explanation an extremely powerful tool for understanding.

Geomorphology overcame the scale problem of the Davisian model by sidestepping it. As Haines-Young and Petch (1986) noted, the solution was not a new theory, it was an alternative means of interpreting the information that already existed. The new interpretation could not be tested directly, but its explanatory framework offered a means of ignoring the temporal and spatial scales of Davisian focus. Schumm and Lichty's classic paper (1965) overcame the Davisian framework by restricting its scope of operation to a specific type of time: cyclical time. This was the longest temporal scale they envisaged.

At this scale only certain variables, as identified by Davis, were of significance in explaining landscape development. In other words, there were only a limited number of variables that changed at this scale.

There were, however, two other temporal scales: steady-state and dynamic equilibrium. At these two scales the variables that were significant for long-term landscape development could be treated as if they were constant, and so were unchanging. Other variables, considered insignificant in cyclical time, became variable, and so significant at the other two temporal scales. Initial relief, for example, is significant as an independent variable in cyclical time but irrelevant as a variable at other time-scales. Drainage network morphology is a dependent variable at cyclical and graded time-scales but an independent variable at the steady-state time- scale.

In this manner Schumm and Lichty provided the beginnings of an explanatory framework that retained the Davisian view but which permitted other conceptual frameworks for explanation to be applied at other temporal, and, by implication, spatial scales. Systems analysis developed rapidly as the integrative explanatory framework of physical geography.

As important, however, were the 'new' questions that this approach made legitimate – questions already being asked and answered about processes of change and their rates. Although the importance of process for understanding geomorphology had been recognized (e.g. Strahler, 1952), and even attempted as a basis for classification in climatic geomorphology, the seemingly immature development of the subject relative to the 'hard' sciences seemed to condemn geomorphology to a 'mere' description of forms.

Identifying and understanding processes and their rates of operation were essential, even to the development of the Davisian explanatory framework. If process rates could be identified from current processes and agents then extrapolation and postdiction could be made and the length of Davisian stages calculated. Changing the scale of study, however, permitted collection of empirical data on processes and so the immediate assessment of the operation of environmental systems.

A spate of publications in the 1970s and early 1980s (Huggett, 1980; Trudgill, 1977; Chorley and Kennedy, 1971) all illustrate the acceptance of process studies as a key feature of geomorphology. This is not to imply, however, that such views were not around before the 1960s and 1970s. Gilbert's work on the Henry Mountains (1877) and dynamic equilibrium was an alternative approach to Davis's historical geomorphology.

The development of the physics of sand movement by Bagnold (1941), the association of mathematical modelling and hydrology by Horton (1945), the study of mass movement by Rapp (1960), as well as the research of Strahler (1952) and Schumm (1956), were all examples of process work in geomorphology. Gregory (1985) assesses the publication of Fluvial Processes in Geomorphology (Leopold et al., 1964) as the most influential publication in the 'new' field of process studies.

These authors were working within an academic context where process studies were not unknown and where systems analysis was beginning to take hold as an alternative explanatory framework to the cycle of erosion. Their work was important, but it illustrates a trend rather than marks a discontinuity. It could also be argued that Gregory was talking about an academic audience that was relatively small and insular.

The textbooks mentioned above were aimed at specialist undergraduate courses, presumably taught by academics who agreed with the approach taken. As important, if not more so, are texts that were written at the time for a wider audience. In this context Cooke's book Geomorphological Hazards in Los Angeles (1984), as well as Cooke and Doornkamp's Geomorphology in Environmental Management (1974), were early examples of attempts to apply concepts and practices in physical geography to an audience beyond the academic.

Indeed, the applied work of Brunsden, Cooke, Doornkamp and Goudie within the UK in the 1970s and 1980s in geomorphology could be viewed as an attempt to develop an applied niche for geomorphological analysis within civil engineering projects and environmental management in general.

By playing to this wider audience, it could be argued that these publications are of more significance in promoting the working practices of physical geography than are the textbooks mentioned above. Similar arguments could be made for US counterparts in applied work such as Graf and Schumm, just to name two. In this manner these workers were applying practical geographic knowledge in much the same way as their nineteenth-century counterparts.

In addition, the significance of graduates pursuing a non-academic career has rarely been considered in subject histories. Graduates who take up posts in consultancy or in management are in positions to influence greatly the type of physical geography applied to 'real' world problems. The trajectory of such individuals, and their use of the academic tools they learnt, has not really been discussed within the literature of physical geography.

This may partly be due to the perception of physical geography as a non- vocation subject. The potential impact of these key figures in promoting and implementing physical geography practices could have aided, via feedback to academics, a more policy orientated approach to the subject with all the baggage of moral and ethical dilemmas that this entails.

The development of a process orientation to physical geography meant that the definition of the discipline itself began to change. The need to reach into other disciplines for the necessary theoretical, field and increasingly mathematical techniques for analysis grew as each subdiscipline became more acquainted with the new focus.

Gregory argues that many individuals became familiar with the new focus through specific textbooks associated with each subdiscipline; for example, desert geomorphology via Cooke and Warren (1973); Embleton and King (1975a, 1975b) for periglacial and glacial landforms; and Carson and Kirkby (1972) for hillslope forms.

This meant that the core of study of physical geography became harder to define. The quest for an integrative explanatory framework for physical geography became somewhat superseded by a search for disciplinary homes with their own, existing frameworks. Whereas under the cycle of erosion the reconstruction of the environment was the core of study once process became the focus of study, each subdiscipline became concerned with process agents.

Studying these could involve the analysis of mathematical models of slopes, or the detailed physics of wind motion to the habits of limpets in weathering rocks. The common focus of study disappeared in the mêlée of process agents. Although it could be argued that the spatial variation of these agents gave each study its geographical flavour, it could equally be argued that the geographer only added an additional viewpoint to an existing field. This problem of discipline definition is a matter that physical geography still has not been able to answer.

Important to the development of a process basis for physical geography was the development of techniques for quantifying landforms and the landscape. The trend to increasingly complex representations of the physical environment within each subdiscipline further added to the separation of the subject. Within climatology and ecology, this trend for making description numerical had long been under way. Climatology had established its data-gathering credentials in the nineteenth century and had encountered and dealt with, albeit in its own way, problems of representation and scale, long before geomorphologists began to grapple with these issues. Likewise, ecology had developed techniques to identify and even simply count the entities of interest at a range of scales by the time geomorphologists got around to considering the problem of what to measure. Quantification of the landscape, or rather components of the landscape, was essential to enable assessment of the contribution of specific processes to the operation of the landscape system.

Despite the great advances made in technology, data collection and a re-orientation towards process studies, physical geography in this period still retained the roots of its past. Despite the new theoretical framework of plate tectonics, physical geography, or rather geomorphology, still retained a focus on landscape development, but at a more human scale.

This promoted relevant studies of landforms and landforming processes within such features as landslides and, more generally, in environment management. Despite the greater volume of information, the push for relevancy and the focus on process studies, the basic tenets of physical geography remained the same. The search for universality, the emphasis on the empirical and a concern with change, albeit as equilibrium and process-response rather than as Davisian stage, remained at the focus of work in physical geography.

The perennial problem of scale remained as a sticking point for the integration of processes found at different scales and acted as a brake on a purely reductionist view of the scientific endeavour in physical geography. A further trend that has been identified by several authors (e.g. Frodeman, 1996; Baker, 1999; Rhoads and Thorn, 1996c; Lane and Richards, 1997; Lane, 2001) is an increasing recognition of the importance of philosophy for the practice of physical geography. Although this may not be a surprise, given the purpose of this book, the manner in which this concern is reflected in the work of physical geographers should not be expected to be consistent. The new concern with philosophy does not reflect a new search for a universal framework for explanation, but rather a questioning of the ability of physical geographers to reach such a consensus.

One outcome of this new concern has been the recognition of the importance of studying small areas in great detail; for example, the significance of small-scale studies as noted by Lane (2001) in relation to fluvial geomorphology. Focusing at this scale of local detail, the universal and contingent become intertwined. Explanation and prediction become more difficult, but also more interesting, as the application of a general solution is not possible without the 'local' also entering into the equation.

2

FUTURE OF PHYSICAL GEOGRAPHY

The following list describes some of the important future trends in physical geography research:

- Continued development of applied physical geography for the analysis and correction of human-induced environmental problems. A student of applied physical geography uses theoretical information from the field of physical geography to manage and solve problems related to natural phenomena found in the real world.
- Remote Sensing - Advances in technology have caused the development of many new instruments for the monitoring of the Earth's resources and environment from airborne and space platforms. The most familiar use of remote sensing technology is to monitor the Earth's weather for forecasting. See the following links for examples of remotely sensed images:
    - Google Maps
    - Google Earth
    - France - SPOT Image
    - USA - Landsat
    - USA - MODIS
    - USA - ASTER
    - Canada - Radarsat
    - GOES
    - USA EROS Data Center
- Geographic Information Systems - A geographic information system (GIS) merges information in a computer database with spatial coordinates on a digital

map. Geographic information dystems are becoming increasingly more important for the management of resources.

Physical geography is done as much by boots and compass as by mental activity. Philosophy in physical geography is an active process open to change as the subject is practised. This book taps into philosophy in action in physical geography by exploring the links between the two and their relations to scientific thought as a whole. Recent developments in geography have highlighted the need for a source of information on the philosophical basis of geography. Texts on the philosophical content of human geography are relatively common (Gregory, 1978; Kobayashi and Mackenzie, 1989; Cloke et al., 1991). Texts that serve the same purpose for physical geography are thinner on the ground.

Apart from the classic Haines-Young and Petch (1986) text on physical geography, there is little accessible to undergraduates that discuss how physical geographers think about and use philosophy within their work. Some physical geographers may regard this as a good thing, automatically stating the Chorley (1978) quote about reaching for their soil auger when they hear the term 'philosophy' (although the rest of the text beyond this infamous quotation is in favour of a more thoughtful approach to the subject). Physical geography has been and still is about 'doing' rather than theorizing in the passive sense of 'armchair' geography, an attitude with which many physical geographers interpret the term. This book is not intending to dispute the active and reflective nature of physical geography; in fact, this book is only possible because physical geography is a subject that views itself as being first and foremost about practice.

The activities of physical geographers, the practice of their subject, involves a vast array of philosophical decisions, even if most are implicit or seen as part of the tradition or training associated with a particular field. Importantly, physical geographers, as all good scientists, question what they are looking at; they do not necessarily take the world for granted as it is. They define, they classify, they probe, they question and they analyse. Why is this not seen as philosophical? The lack of explicit recognition of the importance of philosophy in physical geography may arise from its practitioners' reflection on what their human colleagues regard as philosophical. Human geography seems to have different underlying philosophies – seemingly many more than physical geography.

Human geographers seem to be forever embroiled in the latest debate about the nature of their reality, the construction of worlds, structuralist versus post-structuralist thinking, as well as the post-modernists' debates on the nature of geography. By extension, the number of books addressing explicitly philosophical issues in subdisciplines of human geography are legion by comparison to physical geography. Each faction seems to have its own clique and its own literature. Physical geographers are caricatured as having only one philosophy: the scientific approach. By implication, such an outdated and simplistic philosophical outlook cannot compete with the philosophical sophistication of human geography.

Physical geographers have not necessarily done much to correct this misconception of the scientific approach. Many physical geographers have preferred to retreat into their subject matter, dealing with the detail of their dating methods or the representativeness of their sampling methods, leaving the philosophizing to those who do nothing practical.

There has, however, been a set of wide-ranging philosophical debates that have occurred and are occurring over the nature of scientific investigation. Most physical geographers have been reluctant to enter into this debate in earnest. The 'physics envy' identified by Massey (1999) implies that only the hard sciences can develop philosophies which the 'softer' environmental sciences of geology and physical geography will then dutifully pick up and operate with. Fortunately, recent debates in physical geography with participants such as Frodeman (1995, 1996), Baker (1999), Demeritt (1996, 2001a), Rhoads and Thorn (1996a) and Richards (1990) have shown that physical geographers and their 'practical' approach to the environment have as much to contribute to the developing philosophy of science as do quantum physicists.

Given the highly detached and impractical nature of theoretical quantum physics, it could even be argued that physical geographers can contribute more to the debate operating, as they do, within a discipline that has the physical environment at a more human scale as its subject matter.

Physical geographers have not been aphilosophical, but rather philosophy-shy. At the undergraduate level, the glaring exception to this has been Haines-Young and Petch's (1986) Physical Geography: Its Nature and Methods. Published in the mid-1980s and with an agenda that favoured critical rationalism as the scientific approach, this book provides a very general and readable introduction to some of the key philosophical questions that physical geographers need to address. At a more advanced level is von Engelhardt and Zimmerman's (1988) Theory of Earth Science. This advanced text gave a very Western European view of the philosophical basis of the earth sciences, and covers much of the same ground as the current text. The detailed level of the argument, however, was more complicated than might be expected of undergraduates unfamiliar with the idea of a philosophy in physical geography. More recent publications have taken a historical view of the development of the subject (e.g. Gregory, 1985) or have been collections of papers such as Rhoads and Thorn's The Scientific Nature of Geomorphology (1996c). The latter publication provides an interesting and thoughtful introduction to thinking in geomorphology appropriate for the postgraduate level. It does not, however, provide a coherent and integrated text that explores the recent debates about philosophy in physical geography.

In the nearly two decades since Haines-Young and Petch published their text, the nature of philosophy in science and, in particular, in environmental science has been debated and changed. Critical rationalism still stands as a beacon for understanding reality, but the nature of that reality has been subject to debate, as has been how environmental scientists should, or do, explore and explain it. The caricature of science as a monolithic philosophy with a rigid and singular approach to its subject matter is one that no longer stands up to scrutiny.

In its place is a series of philosophies that share a belief in an external reality. Where these philosophies differ is in their claims about the ability of researchers to obtain an absolute knowledge of reality. Critical rationalism, critical realism and pragmatism have all been flagged in the scientific and geographic literature as being of relevance. In addition, the development of a more sociological view of scientific practice, as illustrated, for example, by the studies of the sociology of science from the Edinburgh school, has not really been addressed within physical geography. Likewise, the potential import of ideas such as Actor Network Theory (ANT) for understanding the context of practice within physical geography is relatively underdeveloped.

One of the things that has become clear is that it is the practice of a discipline that aids the development of its philosophy. Chorley may have been right to reach for his soil auger, but it is only because by the practice of a discipline that anyone can hope to understand its philosophy and, in turn, to understand how and why it changes.

IDEAS, CHANGE AND STABILITY IN PHYSICAL GEOGRAPHY

Ideas are central to understanding how physical geographers have observed and analysed the physical world. Ideas are not static, they change over time as well as from place to place. Ideas influence what we believe and how we understand reality. Ideas change internally, through the practice of a discipline, as well as externally through the social contexts within which physical geographers work. Differing ideologies, different philosophical approaches can influence the questions asked as well as the answers sought. Exploring this complex tangle of influences has often involved detailing the history of a subject. By following the 'great men' (and it usually was men in nineteenth-century academia) and their works, the key concepts from different periods can be traced and some sort of model of change developed. Often as important is the use of such a history to justify the current set of concepts within a subject.

This chapter briefly explores the history of physical geography to try to determine if there is a coherent current, accepted set of ideas that mould geographic thought. There may be no inevitability about the suite of ideas that form the focus of contemporary work. These ideas do not exist in isolation from particular philosophical approaches that have developed or been imported into physical geography. Philosophies can hide in the background, directing and constraining the type of questions asked, and the way in which they may be answered. It is important to provide at least some illumination of these philosophical approaches and their relative importance at different times in and for different parts of physical geography.

What are Ideas and how do they Change?

Ideas and theories are frameworks of thought with which people try to understand physical reality. A more detailed definition of theory will be covered in Chapter 2, but for this chapter an appropriate definition of a theory would be the core ideas or concepts that go unquestioned when a researcher studies the physical environment. From this core flow all the ideas and hypotheses that a researcher has about reality. The core ideas, and the central relations between these and derived ideas, form a tight, logical network. The dominance of vertical movement of land surfaces could be seen as an unquestioned assumption about how reality was in geology at the turn of the twentieth century. From this central, core idea other ideas flowed that explained the distribution of species that did not contradict the evidence on distributions available. Rapidly rising and submerging land bridges were mentally constructed to explain the range of distributions of fossil creatures between South America and Africa. As this central idea was replaced by the acceptance of horizontal movement, continental drift and plate tectonics were developed as an alternative explanation for these distributions.

A number of authors have tried to generalize how ideas change. Three 'big' ideas are progressive change, paradigms (Kuhn, 1962, 1977) and research programmes (Lakatos, 1970). Progressive change states that ideas in science change incrementally and slowly. Current researchers build upon the work of others, correcting their errors until finally, at some vastly future unknown date, everything is understood and reality, as it really is, is known. Accumulation of establishable facts counts as progress of knowledge. Once a fact has been discovered it remains the same for all time. Current research is just adding to and extending existing knowledge. Unfortunately, understanding of reality does not seem to follow a nice, steady and progressive path. There are differences in the rate at which ideas change. Some may suddenly appear apparently from nowhere and some seem to disappear, never to be mentioned in polite scientific circles again, such as racial superiority and acclimatization. This view assumes that earlier researchers understood reality and derived the fundamental laws and relationships that later workers have merely embellished.

Research into critical questions posed to undermine a paradigm was contrasted with the mundane 'normal' science of the established order. But does this romantic view of scientific progress match what actual happens? Lakatos thought that Kuhn's ideas did not match his experience of science and how it changes. Lakatos was a strong supporter of a rational basis for choosing between competing sets of ideas and searched for a way to incorporate a critical rationalist approach into this process of choice. Lakatos divided sets of ideas into research programmes. Each programme had a central set of core ideas, the heart and soul of a set of theories. These directed much of the work the researcher undertook, constraining what questions to ask, how to ask them, and the techniques and explanations that were valid in any answer (the positive heuristic).

In addition, the research programme also identified the questions that researchers should not ask, techniques and modes of explanation that were unacceptable (the negative heuristic). The core theories were never directly tested; they were kept away from direct assessment. Instead a band of auxiliary hypotheses formed the basis of the active use of each research programme. These were testable statements derived from the core theories. These were the elements of the research programme that researchers used in their scientific lives. Disproving one hypothesis did not bring the whole of the research programme into question.

The nodes represent axioms in the core and hypotheses in the outer circle. These nodes are linked by vertices that represent relationships between nodes. The progressive research programme has a rich and internally highly connected core of axioms. These generate a range of hypotheses that can even interact with each other to generate 'second'-generation hypotheses (or more). The degenerative research programme has a limited set of core axioms that generate few isolated hypotheses. That hypothesis was rejected and some other hypothesis constructed using the existing central theories.

Lakatos suggested that choosing between different research programmes became a matter of rational choice. Research programmes that continually generated new hypotheses, novel ways of questioning reality and which were generally not disproved, were progressive programmes. Research programmes that were stagnant and did not advance new ideas or explanations for reality were seen as degenerative. When there were two competing research programmes researchers switched to the more progressive research programme. Some individuals might take longer than others to see that one programme was more progressive, but eventually all would make the same rational choice. In Lakatos's view of science there was elimination of errors and change of world-view, but each change brought the researcher, or rather the scientific community, closer to the truth, to the way reality really is. Unfortunately, it would never be possible to prove that reality had been captured absolutely by the theory, only that the theories of a particular programme had not been disproved.

Both paradigms and programmes have been used to describe changing concepts in physical geography. Neither is perfect and each has a number of problems. Adhering to paradigms would imply that at any one time a single, all-embracing set of ideas would direct the researchers in a subject. This is rarely the case. Similarly, adherence to research programmes implies that central sets of concepts are somehow immune from testing and unaffected by the failure of hypotheses derived from them. Such an ivory tower existence for privileged theories is rarely sustained. Some authors have suggested that there are general currents in thought and that theories follow these. Researchers are moved by fads and fashions as much as anyone else and the flavour of the month has as much relevance in describing academic work as any other. Identifying how such fads are translated to theories or a mechanism by which changing ideas occur within fads has not really been tackled. A significant problem with both frameworks is that they do not really address how and why certain ideas seem to survive as general, favoured theories change, whilst other ideas are rejected. The stability of some concepts, the recognized, but malleable nature of others and the outright rejection of still others is not addressed.

How ideas change is still open to debate despite the models presented here. One of the important points to bear in mind is how each model views the ultimate goal of research. If you believe that knowledge is progressive and developing towards a final complete understanding of reality as it is, then Kuhn's paradigms, with their sociological basis, are unlikely to be attractive. If you believe that all knowledge is relative then paradigms may be more attractive. Lakatos seems to sit on the fence by accepting that ultimate absolute knowledge is unattainable, but then implies that there is a progressive march to this goal that helps to decide between research programmes. This would suggest an underlying belief in the movement of ideas towards representing and explaining reality as it really is, even if you can never be certain you have arrived.

Is there a History of Ideas in Physical Geography?

Physical geography is a diverse subject with many, shifting subdisciplines vying for status. Outlining a single history of such diversity is both problematic and misleading. Detailed histories of ideas and their development can fall into the trap of reading the past by the light of the present.

Trends, unbeknown to workers at the time, are identified with contemporary certainty and clarity. Such histories can lapse into justifying the present rather than trying to understand the past concepts in their appropriate contexts. Often these histories are involved in justifying current practice or concepts by referring to a long and hitherto neglected (or unknown) history. Such reinterpretations of the past often provide intellectual justification for the present by an appeal to the authority figures of a noble past.

The first thing to note in looking at the history of physical geography is that there is not a single history of physical geography. There are many possible histories that can be constructed from the information available.

Judgement of a single narrative as the 'correct' one implies a very restricted view of what is important in physical geography and how it came to be. If there are many possible histories, how can anyone identify what is important from the past? It is essential to have a clear idea of why you want to look at the history of physical geography.

This objective will determine what you consider to be important. In this book, my purpose in reviewing the history of physical geography is not to provide a definite history. Instead there are three key concerns that I want to highlight as being present in the study of the physical environment: universality of explanation, a concern with the empirical, and a concern with explaining change or stability.

Tracing how these concerns have been expressed and the strategies for answering them is the focus of my thumbnail history. Even this limited ambition is fraught with difficulties.

Delving into the history of physical geography requires identifying the subject in the past. This can be a major problem.

To identify individuals, and what they did as physical geography, requires that physical geography itself be seen as a legitimate subject area. As an academic subject, physical geography requires professional academics, academic institutions within which they work, journals within which papers in the field can be published, and so on. The subject requires extensive legitimizing networks of relations between academics, between institutions and other organizations such as publishers. These networks were not necessarily around in the past nor in a form that would permit the identification of physical geography as it is understood today.

A more fruitful approach might be in identifying the central subject matter, and defining the histories of physical geography that way. There is a problem with this approach as well. It assumes that the same subject matter has been studied throughout academic disciplines over time and space. Although it may be possible to identify the ancient Greeks as studying climate, were the phenomena they studied the same as contemporary or even Victorian academics would consider climate? Were they collecting information on the same phenomena?

Even if there was some commonality in the nature of the phenomena under investigation, is it possible to abstract it from the whole intellectual framework of which it is a part? Newtonian mechanics is usually viewed as a key element of the foundation of modern science, and Newton's 'modern' approach of experiment confirmation is viewed as exemplary. The alchemy aspect of Newton's work, and the manner in which this was intertwined with his 'acceptable' science, is rarely mentioned (White, 1998). For Newton, both aspects of his work were vital; he made no distinction between what we would deem 'proper' science and 'pseudo-science'.

Another possible approach is to look at the 'great' figures in the subject and define your histories by reference to what they thought and did. Although it may not be possible to identify them as physical geographers (remember the term might not have existed), at least their work survives to be reinterpreted as the appropriate subject matter of contemporary physical geography.

Appropriating figures such as Darwin, Wegener and Huxley as influential in developing ideas in physical geography neglects the multidisciplinary or even interdisciplinary nature of their work. The environment these individuals worked within was different from the present and the subject areas they recognized would not have necessarily included what contemporaries would view as the 'core' of physical geography. It is also likely that other disciplines will claim these figures as originators in their subjects as well.

Appropriation of 'great' men will involve teasing their relevant work from the rest of their intellectual endeavours, a division they would not have made. It could be argued that even this approach creates histories that mirror contemporary concerns, rather than really reflect the work of these individuals.

Likewise, focusing on great men (and it usually is men in the gender-closed Victorian intellectual world of geography) neglects the contribution of 'lesser' figures who undertook practical work that might be described as geographical. The army of surveyors, for example, who undertook the surveying of the British Empire, trained and developed methods for describing and collating spatially referenced information (Driver, 1992; Collier and Inkpen, 2003).

The data and analysis undertaken by these forgotten figures were vital in developing imperial policies and developing research methods, yet it is often regarded as a lesser achievement than abstract theorizing. Contemporary intellectual snobbery can cloud what is deemed relevant and what is deemed unworthy for a subject's history.

Most Anglophone histories of physical geography tend to focus on the Anglo-American experience and definition of the subject. This limits the types of histories that can be written. Although it could be argued that this impression reflects an inability to read well in other languages, it also reflects a belief in these areas that they represent 'real' geographic thought.

Some histories give nods to early work by Portuguese navigators or Arab intellectuals, but from the eighteenth century onwards the Enlightenment and scientific revolutions bring with them a view of 'proper' science finally being done.

For 'proper', read 'science' as we Western Europeans would understand it today. Such views are also translated into contemporary global surveys of geography. AngloAmerican-derived topics and ideas occupy centre stage; other cultures are relegated to minor interests. It could be argued that this concern with Anglo-American physical geography reflects a wider pattern of change since the eighteenth century.

Development of a capitalist world economic system, some would argue, has marginalized and stifled any independent intellectual development of indigenous cultures. It may be, therefore, not surprising that the only 'acceptable' histories are those with a Western, and specifically an Anglo-American, flavour. Although acceptable to whom – and why them – is another loaded question that is too complicated for consideration here. The possible bias in the construction of histories, and the reasons for it, are rarely touched upon in histories of physical geography.

All the above might imply that there is no point to trying to even produce a historical view of physical geography. There is no single history, there is always the problem of interpretation and there is always likely to be subjective bias.

There is, however, still a case to be made for sketching even a flawed history of physical geography. Identifying ideas within their context and tracing how some ideas have evolved may help in understanding why some ideas remain relatively stable and others fade into obscurity. Stable ideas are those that appear to retain their basic tenets and produce results that are acceptable to the academic community of the time.

The Davisian cycle of erosion, for example, could be viewed as an idea that no longer provides acceptable results upon which contemporary geomorphology can operate. It does not provide a framework for asking or answering questions felt to be appropriate to practising geomorphologists. The cycle of erosion becomes a distinguished or obscure, depending on your viewpoint, element of the history of the discipline.

Plate tectonics, on the other hand, has evolved to be a major and overarching framework for posing and answering geomorphic questions about long-term landscape change. From the 1960s, the general idea of plate tectonics has been developed and refined to create a complicated range of subdisciplines, each with their own particular criteria for evidence and working practices. Tracing the decline of one idea and the rise of another brings to the fore both the shifting nature of what is viewed as real and the surprising stability of much of the content of physical reality.

What are the Important Concepts in Physical Geography?

The three themes developed as central to the history of physical geography are not independent. The quest for universality in explanation has been driven by the need to explain landscapes and their changes. Likewise, identification of the changing or stable nature of the physical environment has been aided by increasingly complex empirical information.

The search for universality can be viewed in terms of both an overarching theory as well as an integrated method for analysing the physical environment. Huxley's Physiography (1877) is viewed by Stoddart (1975) as an important exercise in forming an integrated approach to the subject matter of the physical environment.

Stoddart sees Huxley's approach as being in a direct lineage from the conception of the physical environment of Kant and Humboldt nearly a century before. Physiography starts at the micro-scale, with the familiar, and works outwards to the macro-scale and unfamiliar.

In so doing, Huxley develops an explanatory framework that begins with identifying and classifying causes at the local level and then working from these familiar, or, as he would have it, 'commonsense' illustrations to the wider picture. Cause-and-effect relationships are built from individual experience, emphasizing the empirical over the theoretical, but applied to increasingly larger spatial and temporal scales. The local is then viewed as part of this wider context and understanding of the local requires this wider context.

Huxley's approach had much in common with the 'new' geography being developed by Mackinder (1887) in highlighting the importance and significance of integration and synthesis of information, both physical and human. Much as with Mackinder's new geography, the high ambitions of physiography were never really fulfilled in developing physical geography. Davisian geomorphology overtook Huxley's physiography as the main explanatory framework of geomorphology, and the dream of an integrated study of the lithosphere, atmosphere and biosphere was increasingly replaced by specializations.

From the perspective of the early twenty-first century the failure of both Huxley and Mackinder to have the influence that their holistic and integrative approaches should have had, or were supposed to have, is of interest. These two figures were influential in their own right and yet their ideas, that strike several chords of recognition, failed to develop as the core of the subject. This illustrates the first problem in outlining a history of physical geography. Both Huxley and Mackinder are viewed as using a Darwinian framework for their ideas, as being good disciples of Darwinian thought. Does their failure, then, reflect a failure of Darwinian thought to influence physical geography? Looking at Stoddart (1966) the answer would clearly be no. According to Stoddart, Darwin's views of ecosystems, the integrative nature of the environment and the significance of change have permeated geographical thought.

The central core of Darwinian thinking (Gould, 2000) – that of non-directional evolution through random variations in organisms – has had a lesser impact upon geographical thinking. It is only specific interpretations of Darwinian evolution that have affected geography, not the detailed theory as outlined by Darwin in 1859. The first problem, then, is not to fall into the trap of assuming that there was a single or universally accepted interpretation of a potentially integrating concept such as evolution around in the past.

A vital rallying point in the search for universality in explanation, as noted by Goodman (1967), was the concept of uniformitarianism. The term, usually attributed to Charles Lyell (1833), has been paraphrased as the idea that 'the present is the key to the past', the general idea being that by observing the present in terms of forms and processes it should be possible to apply this knowledge to the past. Everything that happens now should have happened in the past. Gould (1965), amongst others, has pointed out the major problems with maintaining a substantive rather than methodological view of uniformitarianism.

Gould (1965) defined substantive uniformitarianism as the concept that the rates of operation of geological processes have been constant or uniform throughout time and space.

Upholding this view would imply, for example, that there has been a constancy in the number and type of events causing the landscape to change, and so in the magnitude and frequency of the processes causing change. This view of uniformity led Lyell to an extreme cyclical view of landscape and biological change (Kennedy, 1992; Gould, 1987) in which he could seriously envisage a cycle to life with extinct species eventually being recreated.

Methodological uniformitarianism referred to the assumptions that underlie any study of the past and present in a historical and empirically based science.

First, the laws that operate to produce change or stasis are assumed to be constant in both space and time. This is a vital assumption as it means that explanations for change and stasis in one location can be applied to other locations and other time periods with confidence that they will operate in a similar manner.

Second, it is assumed that explanation of the present (and indeed the past) does not require any invocation of unknown or unknowable processes. This is the assumption that uniformitarianism safeguards most saliently. At the time of its development, uniformitarianism was a concept set up in opposition to the prevailing attempt to explain all of nature, and more importantly what was not understood about nature, by final reference to an overarching deity. This final source of all explanation could be invoked to explain why the world was as it was; there was no need to seek a material, earth-based explanation – reference to the guiding principles of a deity was sufficient.

This should not, however, be interpreted as meaning that the explanations offered were simplistic or unreasoned. On the contrary, the most learned brains of the time subscribed to this view and debated the 'correct' interpretation of evidence in a logical and reasoned manner. It is just that the basis of their logical system permitted explanation by reference to a deity as much as explanation by reference to observable processes and agents. Hutton and Lyell used purely material, empirical observations and reasoning restricted to what they could observe and infer about material reality to construct explanations of features in the landscape.

This was a radical step and at a stroke it removed catastrophism from scientific explanation. This was important as it meant that dramatic biblical events, such as the flood, could no longer be used to explain observations such as extinctions or the unconformity of sediments. Likewise, it meant that observations no longer had to be shoe-horned into a temporal sequence that was restricted by biblical events.

The empirical 'facts' now had greater weight in a scientific argument than the enforced explanatory framework. Time, as the arena within which processes operated, was no longer limited to the scale of biblical events. Out of this release, however, another constraint was developed: the idea that there could be no sudden changes in process rates or the occurrence of processes not currently active. This meant that explanations that made use of what would today be considered as extreme events, such as meteorite impacts, were excluded from the explanatory framework.

The important ideas of uniformitarianism lie in the concepts from which it is derived. Uniformitarianism assumes that everything observed in nature can be explained, the principle of causality as an overriding universal principle enters into explanation. Everything observed is capable of explanation, everything observed has a cause.

The cause is not based in some unobservable and hypothetical other world, in some omnipotent deity. Causes are to be found in what we can observe in the here and now. Cause and effect only have recourse to the material world, not to the supernatural. Uniformitarianism is in this sense, as Goodman (1967) suggests, the simplest explanation for the physical world and how it changes. From this one idea, applied consistently and continuously in physical geography, the importance of the empirical rather than the theoretical, of the observable rather than the unobservable, and of the location of causality in the material present are to be found.

Of vital importance in nineteenth- and early twentieth-century thinking about the physical environment was the use, or rather interpretation, of the concept of Darwinian evolution. Darwinian evolution could be seen as representing the application of uniformitarian principles to the organic realm. In this manner, it is another attempt to provide an overarching theory to explain the natural world. Evolution is, however, more than just Lyell amongst flora and fauna.

Bowler (1983a, 1983b) has written on the development of the concept of evolution before and after Darwin, and Livingstone (1984) has illustrated how the concept of evolution was interpreted in a particular manner by American physical scientists.

Darwin had expounded a theory of change in the organic realm by means of random variations in individuals. Individuals competed for resources and, where the random variation was beneficial in this competition, successful individuals produced more offspring. The offspring had the beneficial variation that became a dominant characteristic in the species over time. From initially small differences between individuals multiplied over the vastness of geologic time, species became differentiated. The random variation upon which this process of natural selection was based was non-directional. This means that the course evolution takes is not predictable; uncertainty and randomness lie at the heart of Darwinian evolution.

Central to North American interpretations of evolutionary thinking within physical geography was the interpretation of evolution used by W.M. Davis in his cycle of erosion. Essentially Davis used the concept of progressive change through predefined stages as the basis for his view of evolution.

This view was not the one expounded by Darwin involving random variation and non-directionality. Davis's view, however, was firmly within a North American tradition of orthogenesis and progressive change as developed by Hyatt and, within Europe, by Haeckel. Haeckel viewed changes in evolution as occurring by the addition of stages to an existing established sequence. An embryo, in his view, went through all previous evolutionary stages, with a new stage, that of humans, added at the end of the embryo's development.

The same sequence of stages could be observed in embryos of different species, but for the 'lower' species this happened without the addition of more 'advanced' stages. So whilst a human embryo might be expected to go through all stages from fish through amphibians to reptiles to mammals, a fish would not have these additional stages visible in its embryonic development.

Gould (1977) uses the term 'terminal addition' to describe Haeckel's vision of evolution. Gould also notes that this viewpoint was compatible with the Lamarckian view of the inheritance of acquired characteristics and that Haeckel did indeed subscribe to the transfer of such characteristics between generations. Haeckel's views from biology have a resonance with the ideas of Cope (1887) and Hyatt (1897), the latter a palaeontologist working initially under Agassiz in the US. They believed that although there may be a preordained sequence of stages, organisms could acquire characteristics as well and pass these on through generations.

Central to Cope and Hyatt's work was the idea that the stages that organisms went through in their development could be accelerated or slowed down. Progressive evolution saw stages being condensed further and further into the early stages of development, allowing more time for the addition of acquired characteristics at the terminus of development. Retrogressive evolution saw stages being slowed almost to a stop so that stages are lost before development ends. The former they termed the law of acceleration, the latter the law of retardation.

Their significance lies in the possible influence they had upon the development of evolutionary ideas in the US. Rather than the non-directional, random and chancy process of selection for natural variations in characteristics of an organism independently in each organism's lifetime, Cope and Hyatt strove for an almost opposite view of change. Within their scheme, evolution was about movement through a series of predefined stages with acceleration or retardation adding or losing stages through the sequence. This is the form of evolution that should be borne in mind when discussing Davisian evolution and the development of ideas from it.

Any landform and any landscape had to pass through the stages set by Davis in his cycle of erosion (1899). Any landform could be classified by its morphology to a particular stage in this predetermined evolution, through youth, maturity and old age. Explanation was contained within the theory, within the evolutionary model used.

Observation merely confirmed the position of a specific landscape or landform within this scheme. Stage became the explanation rather than what happened in the landscape. Significantly, however, the stage model highlighted the importance of identifying and providing an explanation for stability and change in the landscape. Both types of behaviour could be explained by reference to the same explanatory framework. Change was an essential feature of evolution and stability was illusory, based on the inability of humans to perceive alteration at a time-scale of relevance to the landscape.

These characteristics of the cycle of erosion provided it with a scientific veneer, particularly as it explicitly used evolution, the scientific theory par excellence in the nineteenth century. Bishop (1980) and Haines-Young and Petch (1986) have pointed out the deficiencies of the Davisian model as a scientific theory, but its perception as a source of progress in explanation in physical geography in the nineteenth century, and as a guiding framework for geomorphic thought in that period, is generally not questioned.

Davis's interest was, however, focused at the macro-level only, although his quest for explanatory frameworks admitted to the operation of the same processes, the same models, across all scales as in his work on tornadoes and their similarity to the spiralling of galaxies. In this manner, Davis believed he had found an overarching and universal mode of explanation for phenomena in the physical environment. It would be inappropriate, however, to view nineteenth-century physical geography as just about Davis and geomorphology.

Even at the scale of the landscape, however, the Davisian model was not the only available theory: Penck had also developed a model of landscape development. Penck's model has been reinterpreted for the systems era as an early attempt to apply a coherent systems framework to the diverse factors affecting landscape development.

Although Penck would not have understood his theory in those terms, his use of factors and their relations to define landscape development does seem capable of easy translation (e.g. Thorn, 1988). A significant difference between Penck and Davis was the nature of changing slope form in each theory. In Davis, slopes declined, whilst in Penck slopes were replaced. In slope decline, denudation processes remove material from the top of a slope at a faster rate than they remove material from the base of the slope.

Gradually, the slope as a whole declines in angle. In Penck's theory, the base of the slope becomes covered with debris denuded from the top of the slope. Gradually, the base of the slope is replaced by a debris slope of constant angle.

In the 1950s King proposed a third type of change: slope replacement. In this theory, slope elements remained constant (waxing, waning, free-face and pediment), only the length of the pediment slope increased over time as debris from the other slope elements reached it and extended it. All other slope elements remained at the same angle throughout their existence.

Although differences in the three theories can be partly related to the different locations of their production, as noted by Thorn (1988), the three theories do illustrate a commonality in their use of time. All three are time-dependent; that is, all three presume that change will be progressive over time and the nature of the form of the landscape will change over time. Denudation chronology has been viewed as a mode of explanation directly linked with the Davisian cycle of erosion and often touted as the dominant mode of explanation in physical geography until the late 1950s.

Denudation chronology attempted to reconstruct the past evolution of the landscape from the sequence of denudation surfaces still visible in the landscape. Evidence for this relative dating of land surfaces was to be found in the morphology of the landscape, its sediments, from absolute dating techniques of portions of the landscape and from an understanding of the processes of landscape development (Gregory, 1985).

Any evidence collected was assessed within a framework of stages of development of the landscape. Evidence did not stand alone, but had to be interpreted within this framework. A great deal of the methodology and theoretical debate over the identification and presence of entities such as planation surfaces centred, as Gregory noted (1985), upon a few 'classic' studies such as Wooldridge and Linton's (1933) interpretation of the South Downs. Even the advent of the statistical analysis of data did not initially diminish the appeal of this approach to landscape study, as trend-surface analysis was employed to aid the identification of 'real' surfaces in landscapes.

SUMMARY

The status of these programmes as true representations of reality is problematic and unlikely to be absolute. Although it is not possible to write the definitive history of physical geography, it is possible to pick out trends that seem to permeate much of the history of the study of the physical environment. Three key themes are identified: the search for universality in explanation, the study of stability and change in the physical environment, and the primacy of empirical information to study the environment. The three themes are related and have evolved together, each affecting and encouraging changes in the others. From these themes have developed the notions of a uniformitarian-based approach to processes in the physical environment. Likewise, the rise of Darwinian evolution and evolutionary interpretations of change in the physical environment have developed from the study of contemporary forms and processes. Central to supporting these ideas was the collection and classification of empirical data about natural phenomena. These themes survived into the late twentieth century, but the questions asked altered. Process-based studies and a concern with complexity echo the themes of universality and change, but with increasingly complicated empirical data to analyse the physical environment.

Ideas in physical geography differ over time and space. Development of understanding about reality can follow one of three models: progressive, paradigms or research programmes. A progressively better and truer representation of the world is achieved by building upon the successes of past researchers in the first model. Knowledge is cumulative and towards an absolute understanding of reality as it really is. Paradigms focus a group of researchers about a specific set of ideas about reality. Paradigm choice is as much about social pressures as about logical, objective choice. There can never be certainty that the understanding derived from one paradigm provides a true representation of reality. Confidence that specific paradigms do provide a true representation of reality may be high, but they can never be absolute. Research programmes highlight that researchers do cluster around specific ideas, but their choice of these ideas is based on logical and objective grounds of theory success.

3

SCIENTIFIC METHOD OF PHYSICAL GEOGRAPHY

INTRODUCTION

Science is simply a way of acquiring knowledge about nature and the Universe. To practice science, one must follow a specific universal methodology. The central theme of this methodology is the testing of hypotheses. A hypothesis can be defined as a proposal intended to explain certain facts or observations that has not been formally tested. The overall goal of science is to better comprehend the world around us. Various fields of study, like physics, chemistry, biology, medicine and the earth sciences, have used science exclusively to expand their knowledge base. Science allows its practitioners to acquire knowledge using techniques that are both neutral and unbiased. The broadest, most inclusive goal of science is to understand (see Figure 3a-1). Understanding involves two interconnected processes: explanation and confirmation. Explanation is perhaps the most important basic goal of understanding. Francis Bacon (1561-1626), a 17th century English philosopher, was the first individual to suggest a universal methodology for science. Bacon believed that scientific method required an inductive process of inquiry. Karl Popper later refuted this idea in the 20th century. Popper suggested that science could only be done using a deductive methodology. The next topic (3b) examines Karl Popper's recommended methodology for doing science more closely.

Explanation consists of explaining reality with a system of hypotheses, theories, and laws. Explanation may also relate observed phenomena to a system of empirical formulas, or link them to mechanisms that are hierarchically structured at both higher and lower levels of function. A theory can be defined as a collection of logical ideas that are used to explain something. The process of testing, refining, and re-testing hypotheses constructs theories. The nature of this confirmation process suggests that theories are rarely static.

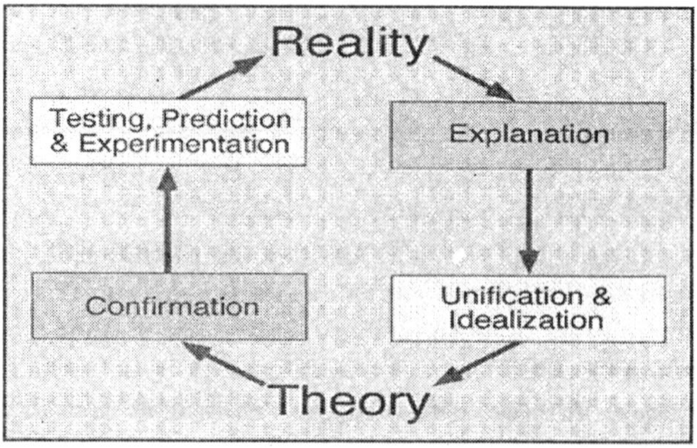

Figure: Facilitating tools involved explanation and confirmation.

Explanation has two important secondary components: idealization and unification. Idealization may be considered to be the condensation of a body of empirical fact into a simple statement. In the process of condensation, some detail must be omitted and the processes and phenomenon abstracted. Idealization may also involve isolating the phenomenon from other aspects of the system of interest. A second aspect of explanation is the unification of apparently unrelated phenomena in the same abstract or ideal system of concepts.

Another minor goal of science is the confirmation of constructed models or theories associated with understanding. Confirmation is accomplished through hypothesis testing, prediction, and by running experiments.

THE HYPOTHETICO-DEDUCTIVE METHOD

Philosopher Karl Popper suggested that it is impossible

to prove a scientific theory true by means of induction, because no amount of evidence assures us that contrary evidence will not be found. Instead, Karl Popper proposed that proper science is accomplished by deduction. Deduction involves the process of falsification. Falsification is a particular specialized aspect of hypothesis testing. It involves stating some output from theory in specific and then finding contrary cases using experiments or observations. The methodology proposed by Popper is commonly known as the hypothetico-deductive method.

Popper's version of scientific method first begins with the postulation of a hypothesis. A hypothesis is an educated guess or a theory that explains some phenomenon. The researcher then tries to prove or test this scientific theory false through prediction or experimentation.A prediction is a forecast or extrapolation from the current state of the system of interest. Predictions are most useful if they can go beyond simple forecast. An experiment is a controlled investigation designed to evaluate the outcomes of causal manipulations on some system of interest.

To get a better understanding of the hypothetico-deductive method, we can examine the following geographic phenomena. In the brackish tidal marshes of the Pacific Coast of British Columbia and Washington, we find that the plants in these communities spatially arrange themselves in zones that are defined by elevation. Near the shoreline plant communities are dominated primarily by a single species known as Scirpus americanus. At higher elevations on the tidal marsh Scirpus americanus disappears and a species called Carex lyngbyei becomes widespread. The following hypothesis has been postulated to explain this unique phenomenon:

The distribution of Scirpus americanus and Carex lyngbyei is controlled by their tolerances to the frequency of tidal flooding. Scirpus americanus is more tolerant of tidal flooding than Carex lyngbyei and as a result it occupies lower elevations on the tidal marsh. However, Scirpus americanus cannot survive in the zone occupied by Carex lyngbyei because not enough flooding occurs. Likewise, Carex lyngbyei is less tolerant of tidal flooding than Scirpus americanus and as a result it occupies higher elevations on the tidal marsh. Carex lyngbyei cannot survive in the zone occupied by Scirpus americanus because too much flooding occurs.

According to Popper, to test this theory a scientist would now have to prove it false. As discussed above this can be done in two general ways: (1) predictive analysis; or (2) by way of experimental manipulation. Each of these methods has been applied to this problem and the results are described below.

Predictive Analysis

If the theory is correct, we should find that in any tidal marsh plant community that contains Scirpus americanus and Carex lyngbyei that the spatial distribution of these two species should be similar in all cases. This is indeed true. However, there could be some other causal factor, besides flooding frequency, that may be responsible for these unique spatial patterns.

Experimental Manipulation

If the two species are transplanted into the zones of the other they should not be able to survive. An actual transplant experiment found that Scirpus americanus can actually grow in the zone occupied by Carex lyngbyei, while Carex lyngbyei could also grow at lower Scirpus sites. However, this growth became less vigorous as the elevation became lower and at a certain elevation it could not grow at all. These results falsify the postulated theory. So the theory must be modified based on the results and tested again.

The process of testing theories in science is endless. Part of this problem is related to the complexity of nature. Any one phenomenon in nature is influenced by numerous factors each having its particular cause and effect. For this reason, one positive test result is not conclusive proof that the phenomenon under study is explained. However, some tests are better than others and provide us with stronger confirmation. These tests usually allow for the isolation of the phenomena from the effects of causal factors. Manipulative experiments tend to be better than tests based on prediction in this respect.

Concepts of Time and Space in Physical Geography

The concepts of time and space are very important for understanding the function of phenomena in the natural world. Time is important to Physical Geographers because the spatial patterns they study can often only be explained in historic terms.

The measurement of time is not absolute. Time is perceived by humans in a relative fashion by using human created units of measurement. Examples of human created units of time are the measurement of seconds, minutes, hours, and days.

Geographers generally conceptualize two types of space. Concrete space represents the real world or environment. Abstract space models reality in a way that distills much of the spatial information contained in the real world. Maps are an excellent example of abstract space. Finally, like time, space is also perceived by humans in a relative fashion by using human created units of measurement.

Both time and space are variable in terms of scale. As such, researchers of natural phenomena must investigate their subjects in the appropriate temporal and/or spatial scales. For example, an investigator studying a forest ecosystem will have to deal with completely different scales of time and space when compared to a researcher examining soil bacteria. The trees that make up a forest generally occupy large tracts of land. For example, the boreal forest occupies millions of hectares in Northern Canada and Eurasia. Temporally, these trees have life spans that can be as long as several hundred years. On the other hand, soil bacteria occupy much smaller spatial areas and have life spans that can be measured in hours and days.

Study of Form or Process

Physical Geography as a science is experiencing a radical change in philosophy. It is changing from a science that was highly descriptive to one that is increasingly experimental and theoretical. This transition represents a strong desire by Physical Geographers to understand the processes that cause the patterns or forms we see in nature.

Before 1950, the main purpose of research in Physical Geography was the description of the natural phenomena. Much of this description involved measurement for the purpose of gaining basic facts dealing with form or spatial appearance. Out of this research Physical Geographers determined such things as: the climatic characteristics for specific locations and regions of the planet; flow rates of rivers; soil characteristics for various locations on the Earth's surface; distribution ranges of plant and animal species; and calculations of the amount of freshwater stored in lakes, glaciers, rivers and the atmosphere. By the beginning of the 20th century Physical Geographers began to examine the descriptive data that was collected, and started to ask questions related to why? Why is the climate of urban environments different from the climate of rural? Why does hail only form in thunderstorms? Why are soils of the world's tropical regions nutrient poor? Why do humid and arid regions of the world experience different levels of erosion?

In Physical Geography, and all other sciences, most questions that deal with why are usually queries about process. Some level of understanding about process can be derived from basic descriptive data. Process is best studied, however, through experimental manipulation and hypothesis testing. By 1950, Physical Geographers were more interested in figuring out process than just collecting descriptive facts about the world. This attitude is even more prevalent today because of our growing need to understand how humans are changing the Earth and its environment.

Finally, as mentioned above, a deeper understanding of process normally requires the use of hypothesis testing, experimental methods, and statistics. As a result, the standard undergraduate and graduate curriculum in Physical Geography exposes students to this type of knowledge so they can better ask the question why?

Descriptive Statistics

Physical Geographers often collect quantitative information about natural phenomena to further knowledge in their field of interest. This collected data is then often analyzed statistically to provide the researcher with impartial and enlightening presentation, summary, and interpretation of the phenomena understudy. The most common statistical analysis performed on data involves the determination of descriptive characteristics like measures of central tendency and dispersion.

It usually is difficult to obtain measurements of all the data available in a particular system of interest. For example, it may be important to determine the average atmospheric pressure found in the center of hurricanes. However, to make a definitive conclusion about a hurricane's central pressure with 100% confidence would require the measuring of all the hurricanes that ever existed on this planet. This type of measurement is called a population parameter. Under normal situations, the determination of population parameters is impossible, and we settle with a subset measure of the population commonly called an estimator. Estimators are determined by taking a representative sample of the population being studied.

Samples are normally taken at random. Random sampling implies that each measurement in the population has an equal chance of being selected as part of the sample. It also ensures that the occurrence of one measurement in a sample in no way influences the selection of another. Sampling methods are biased if the recording of some influences the recording of others or if some members of the population are more likely to be recorded than others.

4

FEATURES OF PHYSICAL GEOGRAPHY

Geographical Set Up: India takes its standard time from the meridian of 82 30 E, which is 5 ½ hours ahead of Greenwich Mean time (0 longitude).Pakistan time is 5 hours ahead of GMT and Bangladesh time is 6 hours ahead of GMT. Significance of Location: Barring the plateau of Baluchistan (which form part of Pakistan), the two great ranges of Sulaiman and Kirthar cut it off from the west. Along the north, the great mountain wall formed by the Hindukush, Karakoram and the Himalayas, which is difficult to cross, cuts it off from the rest of the continent. Similarly, the southward of-shoots of the Eastern Himalayas separate it from Russia. The tropical monsoon climate of India, which ensures a fair supply of moisture and forms the basis of farming in India, is also a result of its location in the southern part of Asia. Since the opening of Suez Canal (1867) India's distance from Europe has been reduced by 7,000km. It thus bridges the space between the highly industrialized nation of the west and the semi-arid, and south-western Asia and the most fertile and populated regions in the south-east and far-east countries.

The Trans Himalayas or Tibetan Himalayas: The largest glaciers are Hispar and Batura (over 57 km long) of Hunza Valley and Biafo and Baltaro (60 km long) of Shigar Valley. The Siachen of Nubra Valley is the longest with a length of over 72 km.The Purvachal or the Eastern Hills: In the east after crossing the Cihang gorge the Himalayas bend towards south forming a series of hills running through Arunachal Pradesh, Nagaland, Manipur, Mizoram, Tripura and eastern Assam and form the boundary between India and Burma.

Physiographic Units of India

S. NO	UNITS	AREA KM (Sq) (appro)	% OF TOTAL AREA
01	Northern mountains	578,000	17.9
02	Great Plains	550,000	17.1
03	Thar Desert	175,000	5.4
04	Central Highlands	336,000	10.4
05	Peninsular Plateaus	1,241,000	38.5
06	Coastal Plains	335,000	10.4
07	Islands	8,300	0.3

Brahmaputra rivers divide Himalayas into three sections: the main Himalayas, the northwest Himalayas and the southeast Himalayas. The main Himalayas running from the Pamir Plateau in the northwest to Arunachal Pradesh in the southeast are the youngest mountains in the world. The highest peak of the word, the Mount Everst(8,884m, named after Sir George Everst). There are about 140 peaks in the Himalayas whose elevation is more than the Mount Blanc (4,810m),the highest peak of the Alps.

The three mountain ranges: the Himadri in the north(the greater Himalayas)the Himachal in the middle(the lesser Himalayas) and the Siwalik(the Outer Himalayas),facing thr palins of India.

The Himadri is of grat elevations (6,000m) which remains covered with everlasting snows. The Siwalik have some flat-floored structural valleys knowns as duns. Dehradun is well-known. Between the Himadri and the Himachal are some broad synclincal valleys. We also classify them as Punjab Himalayas, Kumayun, Assam Nepal and Northern.

Significance of Himalayas: (i) Physical Barriers (ii) Birthplace of Rivers (iii) Influence on climate (iv) Flora and Fauna (v) Mineral Resources (vi) Economic Resources (vii) Tourist

THE GREAT NORTHERN PLAINS

Lies between the great Himalayas in the North and the plateau of Peninsular India in the south. Nearly 2400 km long around 250-320 km broad, the most extensive plan indeed. It is said that this region was once a vast depression, filled with silt – brought down by the three Himalayas River, namely the Indus, the Ganga and the Brahmaputra and their numerous tributaries.

It contains some of the richest soils. The bhangar refers to the upland formed by deposition of older alluvium in the river beds and the Khadar are lowlands formed by deposition of detritus of new alluvium in the river beds.

Bhabar and Terai: Includes those regions where the Himalayas and other hilly regions join the plains. Coarse sand and pebbles are deposited. Bhabar lands are narrower in the east and extensive in the western and north-western hilly. Water converts large areas along the river into swamps known as Terai.

The Western or Rajasthan Plains: are known as Marusthali of Thar and the adjoining Bagar areas to the west of Aravalli. Luni whose water is sweet in the upper reaches and saltish by the time is reaches the sea.

The several salt lakes in the region such as the Sambhar, Degana, Kuchaman and Didwana; from which table salt is obtained. In most of the region shifting sand dunes occur.

The Punjab-Haryana Plain: These plains owe their formation to the Sutlej, the Beas and the Ravi rivers. Many low lying flood plains (called bets) are found here. The Bari Doab between Ravi and Beas rivers, the Bist Doab between the Beas and Sutlej and the Malwa plain are relatively more fertile plain.

The Ganga Plain: The Ganga-Yamuna Doab comprising the Rohilkhand and the Avadh Plain is the tile area that is drained by the tributaries of Ganga.

The North Bengal Plains: The Plains extending from the foot of Eastern Himalayas to the northern limb of Bengal basin cover an area of 23,000km2.

Brahmaputra Plains: This is a low level plain, rarely more than 80km broad, surround by High Mountain on all sides except on the west.

Significance of the Great Plains: Riverine region, Fertile soil, favorable climate, flat surface, constructions of roads, extensive system of irrigation.

The Peninsular Plateau: Oldest structure of the Indian subcontinent whose slow and steady movement towards north and north-east has been responsible for creation of the Himalayas and the Northern Plains in place of the Tethys sea of geological time. It is marked of from the Indo-Gangetic plain by the mountain and hill ranges such as the Vindhyas, the satpura, and Mahadeo, Maikal, and Sarguja ranges with the average height is usually divided into two major subdivisions with the Narmada valley as the line of demarcation. The region north of the Narmada valley is known as the Central Highlands and in south of the Narmada valley lies the Deccan Plateau.

THE CENTRAL HIGHLANDS

Old Aravalli Mountain on the west and the Vindhyas on the south. This region slopes northward to the Ganga plains. The western part of the Central highlands is known as the Malwa Plateau. The Central part has a number of small plateaus like those of Rewa, Baghelkhand and Bundlekhand. The eastern part of the Central Highlands comprises the Chotanagpur plateau.

The Deccan Plateau: The Deccan Plateau extends from the vindhyas to the southern tip of the Peninsula. This triangle plateau is at its widest in the north. The Vindhya Range and its eastern extension namely Mahadev hills Kaimuir Hills and Maikal Range from its northern edge. Western Ghats are known by different regional names such as the Sahyadris in Maharastra and Karnataka, the niligris in Tamil Nadu and Annamalai and the Cardamon hill along the Kerala and Tamil nadu border. The elevation of the ghats increases towards the south. The highest peak, Anaimudi (2,695 m) is in Kerala. The most important gap in the Western Ghats is the Palghat gap which links Tamil Nadu with Kerala. The Bhorghat and the Thalghat are other gaps lying in Maharastra state.

Eastern Ghats: These hills rise steeply from the Coromandel coastal plain. The Eastern Ghats are well developed in the region between the Godavari and Mahanadi rivers. The Eastern Ghats and Western Ghats converge in the Nilgiri hills. Dodda Betta (2,637 m) is the highest peak in the Nilgiri hills.

Significance of Peninsular Plateau: (i) Geological richness (ii) Sources of Irrigation and hydroelectricity (iii) Agricultural Resources (iv) Forest Resources (v) Rich Fauna

THE COASTAL PLAINS AND ISLANDS

The Coastal plains: The Peninsular plateau is bounded by coastal plains on the east and west. There is wide difference between the eastern and western coastal plain. The west coast is narrower but wet. East coast much wider but relatively dry. A number of river deltas occur on the east coast. The deltas of east coast from the 'granary' of the five southern states-Andhra Pradesh, Tamil Nadu, Karnataka, Kerala and Pondicherry. The western coastal strips which have a large number of lagoons and back waters on other hand are noted for spices, areca nuts, coconuts palms etc.

Western coastal Plain: These lies between the Western Ghats and the Arabian seas and stretch from Kutch in the north to Kanyakumari in the south. The Gujarat plain is a broad and flat plain. The Kutch Peninsula, Gulf of Kutch, and the Gulf of Cambay. The Kathiawar Peninsula, also known as Saurashtra which lies to the south of Kutch, is also a plain level area except for some hills rising into Mount Girnar. Sun-divided regionally into the Konkan coast in Maharastra Canara coast in Karnataka and Malabar Coast in Kerala.

Eastern Coastal Plain: The eastern coastal low lands extend from the mouth of the Ganga to Kanyakumari. The northern half is called Northern Circars or Kalinga coast, while the southern half is known as the Coromandal coast. The border part is the Carnatic region, which is about 480 km wide. They are not suitable for harbours as their mouths being full of sit.

Significance of Coastal plains: (i) Harbours (ii) Specialized crops (iii) Fisheries and Navigation (iv) Economic Influence (v) Historical Importance

ISLANDS

Bay of Bengal Islands: The Andaman & Nicobar group of islands are separated by 10 Degree channel. These islands have been formed by extension of tertiary mountain chain of Arakan Yoma. The Andaman group of island is divided into 3 major groups' viz., North Andaman, Middle Andaman and south Andaman, collectively called Great Andamans. Little Andaman is separated from Great Andamans by Duncan passage. The Nicobar group of islands of which Great Nicobar is largest and southern most one. Saddle peak situated in Andaman Islands is the highest peak of Andaman & Nicobar group.

The Dhuliar peak is the highest peak of Nicobar group of islands.More habitable and different origin thanofArabian sea islands. Barran island is dormant volcano and Narcondam island is extinct volcano.

Lakshwadeep islands: Coral- More Muslim population and have fringing reefs. They are 25 small island groups. The island's north is known as Amindivi and spot is Cannonore island. Extreme south is Minicoy islands which is largest.

CLIMATE REGIONS

1. Tropical Rainy climate Region: This region has consistently even temperature, which stays above 18 C even in December, April, and May are the hottest months, the temperature varying from 18 C to 27 C. July and August are the coolest months, which copious rainfall. The average rainfall exceeds 250cm, which encourages wet evergreen forests. The western coastal strip, Western Ghats, south of Bombay, Meghalaya, western Nagaland and Tripura come in this climatic region.
2. Tropical Savanna Region: The chief feature of this climate is the long dry period. Temperature even in

winter stay above 18 C, and in summer may even go up to 46 c. Rainfall, except in the southeastern parts, is in summer and averages about 100cm. In the south eastern parts, the retreating monsoons bring sufficient rains. A major part of the southern peninsula, except the arid tract lying east of the Western Ghats, northeastern Gujarat, south Bihar, major parts of Madhya Pradesh, Orissa, northern Andhra Pradesh, eastern Maharastra, and eastern TamilNadu coast come under this region.

3. Tropical Steppe Region: The average temperature is over 27 C, the lowest temperature being about 23 C April and May are the hottest months, when temperature may rise over 30 C. Average rainfall being less than 75cm, the region comprises a part of the famine zone of the country. The southwest monsoons bring rain to this region. The region comprises the rain shadow areas lying east of the Western Ghats and covers Karnataka, interior TamilNadu, western Andhra Pradesh and central Maharastra,
4. Sub-Tropical Steppe Region: The average temperature rises over 27 C and as high 48 C is recorded during summer. Rainfall from the southwest monsoon averages between 50 to 75 cm. It often fails, leading to widespread drought conditions. This region comprises tracts stretching and encompassing western Rajasthan and northern Gujarat.
5. Tropical Desert Region: The summer Temperature rises over 48 C, while in winter it goes down to 1 C. May and June are the hottest months. Average rainfall is 12.5 cm and very unreliable. Excessive evaporation during summer and intense cold during winter hinder crop production except through river irrigation. Western Rajasthan and parts of Kutch, which are purely sandy plains, come in this region.
6. Humid Sub-Tropical Region: Summer temperatures are 46 C to 48 C. Average rainfall, mostly from summer monsoons, is about 62.5 cm, which increases to over 250cm in the east. Winters are virtually dry. This region

is spread over the foothills of the Himalayas, eastern Rajasthan, plains of U.P, Bihar, northern Bengal, part of Assam and Arunachal Pradesh.

7. Mountain Region: Average temperature in June is 15 C to 17 C, in winter it goes below 8 C. On the northern slopes of the Himalayas, the rainfall is scanty, about 8-10 cm, but the western slopes enjoy a heavy rainfall of over 250 cm. The entire Himalayan belt comprising both the trans-Himalayas and the main Himalayas come under this region.

THE PENINSULAR PLATEAU

The Peninsula plateau is roughly triangular in shape with base coinciding with southern edge of the great edge of the Great Plains. It covers a total area of 160,000km 2. (About half of total land area of the country)

The Plateaus of Peninsular India

1. The Marwara Upland: It lies east of Aravali range made up of stand stone, shales and limestone of Vindhayan period. The upland is drained by Banas River originating in Aravali range.
2. The Central Highland: Also called the Madhya Bharat Pathar, lying in the east of Marwara upland and drained by Chambal and its tributaries.
3. The Bundelkhand Upland: To south of Yamuna river between Madhya Bharat Pathar and the Vindhyan scarap lands consists of granites and gneiss.
4. Malwa Plateau: Bonded by Aravali range in the west and Bundlekhand in east and Vindhayas in south. It is an extensive lava plateau in M.P. with black soil drained by a number of north flowing rivers like Betwa, Parbati, Kali Sindh, Chambal and Mahi.
5. Baghelkhand: East of Makal range and bounded by Son River on the north.
6. Chhotanagpur Plateau: It covers mostly Jharkhand adjoining eastern Madhya Pradesh and Purulia of West Bengal. This Plateau is composed mainly of Gondwana

rocks with patches of granites ad gneisses and Deccan lavas. The highest general elevation is in the Midwest portion known as patlands (high level laterite plateau e.g. Netarhat). The Plateau is drained by numerous rives in a radial pattern such as Damodar, Subarnrekha, North Koel, South Koel and Barakar river.

7. Meghalaya Plateau: This Plateau has been separated from main block of the peninsular plateau by a gap called Ganu-Raj Mahal gap. From east to west the plateau comprises Gana, Khasi and Jaintia hills and Mikir hills. Shilong is the highest point of the plateau.
8. The Deccan Plateau: This is the largest unit covering an area of 5 lakh sq.km. It comprises Maharastra Plateau, Karnataka plateau and the Telengana plateau (Andhra Pradesh). The general slope is form west to east. Indicated by flow of major rivers like Mahanadi, Godwari, Krishna and Cauveri.
9. The Chhattisgarh Plain: It is the only plain worth the name in the vast stretch of plateau drained by upper course of Mahanadi. This saucer shaped depression lies between Makal range and Orissa hills.

HILL RANGE OF THE PENINSULA

- Aravali Range: Aravalis are the world's oldest fold mountain running in north east to southwest direction from Delhi to Palampur in Gujarat. Gurushikhar is the highest peak situated in Mount Abu. Barr, Piplighat, Dewair and Desuri passes allow movements by roads and railways.
- Vindhyan Range: It raises as an escarpment flauting the northern edge of the Naramada, Son-trough. The Vindhyas are continued eastward as the Bharner and Kaimur hills. This range acts a s a water divided between Ganga system and river system of South India. The Makal range forms a connecting link between Vindhyan and Satpuras.
- Satpura Range: It is a series of seven (Sat) mountains running in east-west direction south of vindhyas and in between the Narmada and Tapi. Commencing from

Rajpipla hills in the west through the Mahadeo hills it extends to Makal range in the east. Dhupgarh near Panchmarhi on Mahadeo hills is highest peak of Satpuras.

- Western Ghats (Sahyadris): The western Ghat run in north-south direction from Tapi valley to north of Kanyakumari. Along the Arabian sea-coast for about 1600 Km. these are Block Mountains due to the down warping of the land to the Arabian Sea. Sahyadris form the real water divide of the peninsula. All the important rivers rise from these hills and flow eastward. The Sayadris up to 16 north latitude are mainly composed of basalt. In this portion Kalsubai is the highest peak. South of Goa Sahyadris are composed of granites and Gneisses. In the Nilgiri hills, Eastern Ghats join the Sahyadris to form a mountain knot whose highest point is Doddabetta (2637 m). South of it is Palghat gap connecting Tamil Nadu with Kerala. South of Palghat there is mountain knot comprising Annamalai hills in north. Patni hills in the north-east and the Cardamon hills in the south. Anaimudi (2695 m) situated in the Annamalai hills is the highest peak of Sahyadri. Kodaikanal hill station is located on the Patni hills. There are 3 important passes in the Sahyadris: (a) Thalghat (b) Bhorghat (c) Palghat
- Eastern Ghat: It is a chain of highly broken and detached hills starting from Mahanadi in Orissa to Vaigai in Tamil Nadu. These are part of the very old fold mountains. It is only in the northern part between Mahanadi and Godavari that Eastern Ghats exhibit true mountain character comprising Maliya and Maduguala koṇda ranges, South of Godawari the broken hill ranges are Nallamalai, Palkonda, and Javadi, Shavroy and Biligiri Rangan hills. Mahendragiri is the tallest peak of eastern Ghat.

SOILS

- COMPOSITION OF SOIL: Soil is the loose material which forms the upper layer of the mantel rock, i.e., the layer of loose fragments which covers most of the earth's land area. It has definite and constant composition. It contains both decayed plants and animals substances. The four main constituents of soils are;

(i) Silica: The chief constituent of sand

(ii) Clay: is a mixture of silicates and contains several minerals such as iron, potassium, calcium, sodium and aluminum. Particles of clay absorb water and swell.

(iii) Chalk: (calcium carbonate) provides calcium, the most important element for the growth of plants.

(iv) Humus: is not a mineral, it is an organic matter. It is formed by decomposed plant remains, animal manure and dead animals and is the most important element in the fertility of the soil. It helps retain moisture in the soil and helps the plant in absorbing materials from the soil for building its body. A soil looks dark on account of the presence of humus.

- TOP SOIL AND SUB SOIL: Two layers namely top soil and sub soil. Top soil(the upper layer) is of greater importance. Good top soil means good crops. It varies considerably in depth and also in character and ability to grow crops. It is only a few meters deep. Million of bacteria, insects and worms live in it. Top soils develop very slowly. It may take years to form top soil suitable for plants, but it can be washed away in a few years if proper precautions are not taken. Sub soils consist of the parent material from which soil is formed. It also contains plant food and moisture but it is not a s productive as top soil. It has to be converted into soil and it may take years to convert sub soil into soil. Below the sub soil generally there is solid rock.
- FORMATION OF SOIL: The natural processes involved in the formation of soil are

  (i) Weathering: process of disintegration of rocks into soil.

  (ii) Deposition: process of progressive lying down of rock particles carried by rivers, ice, marine, currents wind or tides.

(iii) Biochemical processes

Soil formation depends upon the following factors; *Nature of the parent rock, * Climate, * Natural Vegetation, * Topography, * Time

- GENERAL CHARACTERISTICS OF DIFFERENT SOILS
  a) Sandy Soil (light soil): It contains more than 60% sand and les than 10% clay. It is easy to cultivate and is favoured for fruits and vegetables.
  b) Clayey Soil: It has high proportion of clay. It becomes sticky when mixed with water. A soil very rich in clay is called 'heavy',
  c) Loam: is rich soil and consist of a mixture of sand and clay. All loamy soils are good for farming and general gardening.
- MAJOR SOIL TYPES: The Indian Council of Agricultural Research (ICAR) classifies the soils of India into eight categories.
- ALLUVIAL SOIL: This is the most important and widespread group of soils. It covers about 15 lakh sq.km of the land area in Great Plains from Punjab to Assam and also in the valleys of the Narmada and Tapi, Mahanadi, Godavari, Krishna and Cauvery. These soils have been brought down and deposited by three great Himalayan rivers Sutlej, Ganga and Brahmaputra. These soils consist of varying proportions of san, silt and clay. Khadar is the newer alluvium which is sandy, light coloured and occurs near river beds where deposition takes place regularly and bhangar or older alluvium. Alluvial soils as a whole are very fertile and therefore the best agricultural soils of the country. The regions of these of these soils constitute the' wheat and rice bowls' of India.
- BLACK SOIL: As the name indicates these soils are black in Colour and since they are ideal for growing cotton, they are also called cotton soil. These soils, covering an area of 5.46 lakh sq.km, are most typical of the Deccan trap (Basalt). The black Colour of regur

is variously attributed to the presence of titaniferrous magnetite, compounds of iron and aluminum-Unsuitable for heavy irrigation. Ideally suited to dry farming. Suitable for cotton, cereals, oilseeds like linseed, castor and safflower, many kinds of vegetables and citrus fruits.

- RED SOIL: These soil occupy about 5-18 lakh sq.km over the peninsula reaching up to Rajmahal Hills in the east, Jhansi in the north and Kutch in the West. In North western peninsula is covered by the black soils and the remaining south-eastern half is covered by red soils- Entire black soil in the eastern part of Peninsula comprising of Chhotanagpur plateau, Orissa, east Madhya Pradesh, Telengana, the Niligris, Tamilnadu plateau and Karnataka. The soil have reddish colour due to iron compounds. It is suitable for rice, ragi, tobacco and vegetables.
- LATERITE SOILS: These soils occupying an area of 1.26 lakh sq.km- Found capping the flat uplands, and are spread in western coastal region receiving very heavy rainfall – Poor in nitrogen, phosphoric acid, potash, lime – suitable for rice, ragi, tapioca and cashewnuts.
- FOREST AND MOUNTAIN SOIL: These soils occupy about 2.85 lakh sq.km in the hilly regions of the country– described as soils in the making – found in the Himalayas and the other ranges in the north and high hill summits in the Sahyadris, Eastern Ghats and Peninsula – poor in potash, phosphorous – Temperate fruits, maize, wheat and barley are grown in Jammu and Kashmir and Himachal Pradesh where soils are mostly podzols which are acidic in re-action.
- ARID AND DESERT SOILS: Northwestern parts of the country and occupy about 1.42 sq. km area in Rajasthan, south Haryana, north Punjab and Rann of Kutch. Thar Desert alone occupies an area of 1.06 sq. km. It contains high percentage of soluble salts and a low to very low organic matter.

- SALINE AND ALKALINE SOILS: These soils occupy about 170 lakh sq. km of arid and semi-arid areas of Rajasthan, Punjab, Haryana, Uttar Pradesh, Bihar and whole of Maharastra. Textually they are sandy to loamy sand. Saline soils contain free sodium. Crops grown on these soils include rice, wheat, cotton, sugarcane and tobacco.
- PEATY AND MARSHY SOILS: These soils cover an area of about 150 sq. km in the Kottayam and Allepey districts of Kerala. They are suitable for paddy cultivation.
- SOIL FERTILITY: The factor responsible for deficiency of Indian soils are: (i) loss of nutrients, largely brought about through the removal of harvested crops, (ii) leaching which occurs under heavy monsoonal rains, causes loss of nutrients, sandy soils are more subject to leaching than the heavier ones and bare soils are more than those covered by plants. Indian soils are mainly deficient in nitrogen, phosphorus and potash. These can be supplied to the soils by application of organic manures and fertilizers.
- SOIL EROSION: Soil erosion is the removal of soil particles by natural agencies such as water and wind and also as a result of human and animal interference.
- TYPES OF SOIL EROSION: Two types of soil erosion:
  (i) Water Erosion: The important types of this erosion are sheet, rill and gully. In sheet erosion thin layer of soil is removed by the water during heavy rains. If the erosion continues unchecked, numerous finger-shaped grooves may develop all over the area as a result of the silt-landen run-off. This is called rill erosion. The rills may deepen and enlarge into gullies.
  (ii) Wind Erosion: Wind erosion is mainly confined to arid and semi-arid areas devoid of vegetation. Wind, particularly during sand-storms, lifts and carries away soil particles leaving behind a depression devoid of top fertile soil. Rajasthan and the adjoining

areas of Haryana, Uttar Pradesh and Gujarat show this type of soil erosion.

- CONSEQUENCEȘ OF SOIL EROSION: (i) heavy floods in rivers, (ii) lowering down of sub soil water level, (iii) reduction of soil fertility, (iv) silting of streams and water courses, (v) disappearance and downfall of civilizations.
- a) SOIL CONVERSATION: Soil conservation is an effort made by man to prevent soil erosion to retain the fertility of soil. It may not be possible to stop soil erosion entirely. b) Measures for Soil Conservation: (i) Planting cover crops, (ii) Adoption of correct farming techniques, (iii) Terracing the practice of cutting steps in hillside, to create level land for cultivation, (iv) Construction of check dams, (v) Creation of wind break, (vi) Controlling of grazing of pastures, (vii) Suspending cultivation for one season and more so as to help the soil recover its fertility.

THE EARTH'S CRUST

The earth is made up of several concentric layers. The outer layer of the earth's crust is called lithosphere, which comprises two distinct parts. The upper parts consist of granitic rocks and form the continents.

Its main mineral constituents are silica and alumina knows as Sial. It has an average density of 2.7. The lower part is a continuous zone of densers, basaltic rocks forming the ocean's floors. Comprising mainly silica, iron and magnesium known as Sima and has an average density of 3.0. The Sial and the Sima together form the earth's crust which varies in thickness 4.8 to 6.4 km, beneath the oceans and about 48 km under some o continents. Beneath the lithosphere is the mantle known as the mesosphere about 2880 kms thick. Composed mainly very dense rock rich in Olivine.

The interior layer is the core known as baryshpere about 2456 km in radius is made up mainly iron (Fe) and nickel (ni) known as nife. The temperature of core is as high as 2000 C and subject to extremely high pressure. The crust forms only

0.5% of the volume of the earth, 83% consist of the mantle and 16% makes the core that of the earth.

The mean radius of the earth is 6400 km. the average rate of increase temp 1 C for every 32 meters of descent. But in upper 100 kms, the increase is estimated at 12 C/kms. It is 2 C/Km in next 300 kms and 1 C/kms below it.

Name of the layer	Chemical Composition	Volume (Earth)	Mars (Earth)	Density	Physical Properties
CRUST (A) (i) Outer-part of lithosphere (ii) Inner part of Lithosphere	Sial	1% of the earth	10%	2.75 to 2.90	Solid
MANTLE (B) (i) Outer part of mantle (ii) Inner part of mantel	Silica with Sima Wholly Sima	83%	68%	3.10 to 4.75 4.75 to 5.00	Partly molten Close to melting point
CORE (C) (i) Outer part of Core (ii) Inner part of Core	NiFe	16%	32%	5.1 to 13	Plastic state Solid and rigid because of tremendous overlying pressure.

Composition of Earth's Crust

1) oxygen: 46.6% 2) Silicon- 27.27% 3) Aluminium: 8.13% 4) Iron- 5% 5) Calcium-3.63% 6) Sodium- 2.83% 7)Pottasium- 2.59% 8) Magnesium- 2.09% 9)Others- 1,41%

CLASSIFICATION OF ROCKS ACCORDING TO ORIGIN

IGNEOUS ROCKS: Igneous rocks are formed by the cooling and solidification of molten rock (magna) from beneath the earth's crust. They one normally crystalline in structure. They do not occur in layers and not contain fossils. Some rocks can form a high percentage of silica are said to be acid. E.g., Granite. Other rocks such as Basalt contain a high percentage of iron or Aluminium or magnesium oxides are called basic rocks. There are two main groups:

- Volcanic Rocks: These are mole nocks poured out of volcanoes as lavas. They solidify rapidly on the earth's surface and the crystals are small. E.g. basalt.
- Plutonic Rocks: These are igneous rocks solidified deep in the earth's crust and they reach the surface only by being exposed by erosion. E.g., granite, diorite and gabbros etc.,

SEDIMENTARY ROCKS: Sedimentary rocks are formed from sediment accumulated over long periods, usually under water and deposited usually in layers by water, wind or moving ice. They are non-crystalline and often contain fossils of animals, plants and other micro-organisms. There are three main groups;

- Mechanically-formed: (a) Wind-deposited e.g loess, (b) River-deposited e.g., clays, gravels, and alluviums, (c) Glaciers-deposited e.g., morains, sands and gravels and boulder clay, (d) Sea-deposited.
- Organically formed: (a) From animals e.g., chalk and coral, (b) From plants e.g., Peat, lignite, coal.
- Chemically formed: e.g., rock slat, borax, gypsum, nitrates, potash and certain limestone.

METAMORPHIC ROCKS: All those rocks whose structure and appearance have been changed by great heat or great pressure or both. Any rock can be changed into a metamorphic rock. In theses manner (a) State (from clay), (b) Marble (from limestone), (c) Quartzite (from sandstone), (d) Graphite (from coal), (e) Gneiss (from granite), (f) Schist (from shale).

STRUCTURE OF ATMOSPHERE

According to Petterssen, the atmosphere is divided into the following five layers:

TROPOSPHERE: The lowermost layer of the atmosphere is known as Troposphere and is the most important layer because almost all of the weather phenomena is occurred in this layer. E.g., Fog, cloud. Thunder, lightning, etc., occurs in this layer. It extends roughly to a height of eight kilometers near the poles and about 18 kilometers at the Equator. The thickness of the troposphere at the equator is the greater heights by strong conventional currents. Temperature decreases with increasing height at the rate of 6.5 C per 1000m. This rate of decrease of temperature is called normal lapse rate. This layer contains dust particles and over 90% of the earth's water vapour. The upper limit of the tropopause is called troposphere which is about 1.5km. Thick. Temperature ceases to fall with the air temperature at the troposphere is about-80 C over the equator and about -45 covers the poles. The word troposphere literally means zone or region of mixing whereas the word tropopause means where the mixing stops.

STRATOSPHERE: The Stratosphere begins at the tropopause which forms its lower boundary. The lower stratosphere is isothermal in character, i.e, the temperature in the lower part of this layer does not change with altitude.

The stratosphere extends up to a height of 50 km. Afterwards it gradually increases up to a height of 50 km because of the presence of ozone layer which absorbs the Sun's Ultra-Violet Rays. Clouds are almost absent and there is very little dust or water Vapours. The air movement is almost horizontal. The upper boundary of the stratosphere is called Statopause. Above this level there is a steep rise in temperature.

MESOSPHERE: Over the stratopause there exists the third layer known as mesosphere. It extends up to a height of 80 kms. Temperature decreases with height again and reaches up to – 100 C at the height of 80 kms.

IONOSPHERE: Ionosphere is located between 80 km to 400 kms. It is electrically charged layer. Radio waves transmitted from the earth are reflected back to the earth by this layer. Temperature again starts increasing with height because of radiation from the sun. The ionosphere consists of the following ionized layers:

D – Layer: 60 -99 km

E – Layer: 90 – 130 km

Sporadic Layer: 110 km

E2 Layer: 150 km

F1 Layer, F2 Layer: 130 -1890 km

G Layer: 400 km and above.

EXOSPHERE: The outermost layer of the earth's atmosphere is known as the exosphere which lies between 400 and 1000kms. The atmosphere in this region is so rarefied that it resembles a nebula-hydrogen and helium gases predominates in this outermost region. The temperature becomes 5568 C at its outer limit but this temperature is entirely different from the air temperature of the earth's surface.

LATITUDE, LONGITUDE AND TIME

The earth is nearly a sphere and it has no edges. It presents some difficulties in positioning its surface features. To over come this problem, a network of imaginary lines is drawn on a globe or a map to help us locate places. The spinning of the earth on its axis provides – the North Pole and South Pole. They form the basis for the geographical grid. The grid consists of two sets of lines – horizontal and vertical. The horizontal lines, running eat-west, are circular and parallel to each other.

The line drawn midway between the North Pole and the South Pole is called the equator. It is the largest circle and hence, is called a great circle. All other parallels get smaller in size in proportion to their distances form the equator towards the Poles. These imaginary lines running east-west are commonly known as parallels of latitude. The vertical lines, running north-south are halves of the great circle and join the two Poles.

They are called meridians of longitude. The latitudes and longitudes are commonly referred to as coordinates because they provide systematic network of lines upon which position of various surface features of the earth can be represented.

LATITUDES: The latitude of a place on the earth's surface its distance north to south of the equator measured along the meridian of that places as an angle from the centre of the earth. Lines joining places with the same latitudes are called parallels. The value of equator is 0 and the latitude of the poles are 90 N and 90 S.

If the parallels of latitude are drawn at an interval of one degree, ther will be 90 parallels in the northern and southern hemisphere each. The total number of parallels thus drawn including the equator, the letter N or S is written along with the value of latitude.

If the earth were a perfect sphere – 23 ½ N –Tropic Cancer, 23 ½ S – Tropic of Capricorn, 60 ½ N – Artic, 60 ½ S – Antarctic, 1 of latitude (a one- degree arc of a meridian) would be a constant value i.e., 111 km – a degree of latitude changes slightly in length from the equator it is 110.6 km, at the poles it is 111.7 km.

Latitude of a place may be determined with the help of the altitude of the sun or the Polar Star.

LONGITUDE: Unlike the parallels of latitude, which are full circles, the meridians of longitude are semi-circles touching the poles. Meridians intersect the equator at right angles.

The meridian of longitude passing through the Greenwich observatory (near London) has been adopted as the Prime Meridian by an international agreement and has been given the value of 0 degree. The numerical value of a longitude varies form 0 degree to 180 degree East or West– be 360 degree meridians.

LONGITUDE AND TIME: While rotating on its axis, the earth completes one circle (360 degree) in approximately 24 hours time. In other words, it move 15 degree longitudes per hour. As such the sun appears to be traversing 15 degree of longitude from east to west every hour or 1 degree of longitude in every 4 minutes.

Thus, when it is 12 noon at Greenwich (Prime meridian), the time at 30 degree east of Greenwich will be 30*4= 120 minutes or 2 hours ahead of Greenwich time. In order to maintain uniformity of time as far as possible within the territorial limits of a country us taken as the standard meridian and its local time is taken as the standard time for the whole country. The standard meridian is selected in a manner that it is divisible by 15 degree or 7 degree 30' so that the difference between its standard time and the Greenwich Mean Time may be expressed as multiples of an hour or half an hour.

The Standard meridian of India is 82degree30' E and the standard time is 5 ½ hours a head of Greenwich Mean Time.

5

EVOLUTION OF THE UNIVERSE

INTRODUCTION

About 11 to 15 billion years ago all of the matter and energy in the Universe was concentrated into an area the size of an atom. At this moment, matter, energy, space and time did not exist. Then suddenly, the Universe began to expand at an incredible rate and matter, energy, space and time came into being (the Big Bang). As the Universe expanded, matter began to coalesce into gas clouds, and then stars and planets. Our solar system formed about 5 billion years ago when the Universe was about 65% of its present size. Today, the Universe continues to expand.

WHY DO MOST SCIENTISTS ACCEPT THE BIG BANG THEORY?

The acceptance of this theory by the scientific community is based on a number of observations. These observations confirm specific predictions of the Big Bang theory. In a previous section, we learned that scientists test their theories through deduction and falsification. Predictions associated with the Big Bang theory that have been tested by this process are:

1. If the Big Bang did occur, all of the objects within the Universe should be moving away from each other. In 1929, Edwin Hubble documented that the galaxies in our Universe are indeed moving away from each other.
2. The Big Bang should have left an "afterglow" from the explosion. In the 1960s, scientists discovered the existence of cosmic background radiation, the so-called

"afterglow" after the Big Bang explosion. Our most accurate measurements of this cosmic radiation came in November 1989, by the Cosmic Background Explorer (COBE) satellite. The measurements from this satellite tested an important prediction of the Big Bang theory. This prediction suggests that the initial explosion that gave birth to the Universe should have created radiation with a spectrum that follows a blackbody curve. The COBE measurements indicated that the spectrum of the cosmic radiation varied from a blackbody curve by only 1%. This level of error is considered insignificant.

3. If the Universe began with a Big Bang, extreme temperatures should have caused 25 percent of the mass of the Universe to become helium. This is exactly what is observed.
4. Matter in the Universe should be distributed homogeneously. Astronomical observations from the Hubble Space Telescope do indicate that matter in the Universe generally has a homogeneous distribution.

How will the Universe End?

Cosmologists have postulated two endings to the Universe. If the Universe is infinite or has no edge, it should continue to expand forever. A Universe that is finite or closed is theorized to collapse when expansion stops because of gravity. The collapse of the Universe ends when all matter and energy is compressed into the high energy, high-density state from which it began. This scenario is of course called the Big Crunch. Some theorists have suggested that the Big Crunch will produce a new Big Bang and the process of an expanding Universe will begin again. This idea is called the oscillating Universe theory.

Early History of the Earth

Scientists believe the Earth began its life about 4.6 billion years ago. The Earth formed as cosmic dust lumped together to form larger and larger particles until 150 million years had passed. At about 4.4 billion years, the young Earth had a mass similar to the mass it has today. The continents probably began forming about 4.2 billion years ago as the Earth continued to cool.

The cooling also resulted in the release of gases from the lithosphere, much of which formed the Earth's early atmosphere. Most of the Earth's early atmosphere was created in the first one million years after solidification (4.4 billion years ago). Carbon dioxide, nitrogen, and water vapor dominated this early atmosphere. Table below describes the three major stages of development of the atmosphere.

As the Earth continued to cool, the water vapor found in the atmosphere condensed to form the oceans and other fresh water bodies on the continents. Oxygen began accumulating in the atmosphere through photo-dissociation of O_2 from water, and by way of photosynthesis (life). The emergence of living organisms was extremely important in the creation of atmospheric oxygen and ozone. Without ozone, life could not exist on land because of harmful ultraviolet radiation.

At about 450 million years ago, there was enough oxygen in the atmosphere to allow for the development of a stratospheric ozone layer that was thick enough to keep terrestrial life protected from ultraviolet radiation.

As a result, terrestrial life began its development and expansion at this time. Table 5b-2 describes the timing of the evolutionary development of some of the Earth's dominant forms of life before and after 450 million years before present (BP).

Table: Approximate origin time of the major plant and animal groups.

Organism Group	Time of Origin
Marine Invertebrates	570 Million Years Ago
Fish	505 Million Years Ago
Land Plants	438 Million Years Ago
Amphibians	408 Million Years Ago
Reptiles	320 Million Years Ago
Mammals	208 Million Years Ago
Flowering Plants (Angiosperms)	140 Million Years Ago

The Natural Spheres

From the standpoint of Physical Geography, the Earth can be seen to be composed of four principal components:

- Lithosphere - describes the solid inorganic portion of the Earth (composed of rocks, minerals and elements). It can be regarded as the outer surface and interior of the solid Earth. On the surface of the Earth, the lithosphere is composed of three main types of rocks:
  - o Igneous - rocks formed by solidification of molten magma.
  - o Sedimentary - rocks formed by the alteration and compression of old rock debris or organic sediments.
  - o Metamorphic - rocks formed by alteration of existing rocks by intense heat or pressure.
- Atmosphere - is the vast gaseous envelope of air that surrounds the Earth. Its boundaries are not easily defined. The atmosphere contains a complex system of gases and suspended particles that behave in many ways like fluids. Many of its constituents are derived from the Earth by way of chemical and biochemical reactions.
- Hydrosphere - describes the waters of the Earth (see the hydrologic cycle). Water exists on the Earth in various stores, including the atmosphere, oceans, lakes, rivers, soils, glaciers, and groundwater. Water moves from one store to another by way of: evaporation, condensation, runoff, precipitation, infiltration and groundwater flow.
- Biosphere - consists of all living things, plant and animal. This zone is characterized by life in profusion, diversity, and ingenious complexity. Cycling of matter in this sphere involves not only metabolic reactions in organisms, but also many abiotic chemical reactions.

All of these spheres are interrelated to each other by dynamic interactions, like biogeochemical cycling, that move and exchange both matter and energy between the four components.

The Gaia Hypothesis

In 1965, J.E. Lovelock published the first scientific paper suggesting the Gaia hypothesis. The Gaia hypothesis states that the temperature and composition of the Earth's surface are actively controlled by life on the planet.

It suggests that if changes in the gas composition, temperature or oxidation state of the Earth are caused by extraterrestial, biological, geological, or other disturbances, life responds to these changes by modifying the abiotic environment through growth and metabolism.

In simplier terms, biological responses tend to regulate the state of the Earth's environment in their favor.

The evidence for Gaia is as follows:

- If not continually replaced by biotic activities gases like methane and hydrogen would become non-existant in the atmosphere in a few decades.
- Carbon dioxide (CO_2) in the Earth's atmosphere is far less abundant than chemistry alone would allow. If life was deleted carbon dioxide would become 30 times more abundant. Large quantities of carbon dioxide are currently locked up by living organisms.
- The Sun's energy output has increased by 30% in the past 3.5 billion years. Yet, historical climate data indicates that the temperature of the Earth has only fluctuated by about 5° Celsius from the current average global temperature of 15° Celsius. Computer climate models suggest that a 30% reduction in solar radiation would create a global average temperature of between -10 and -52° Celsius all things being equal. These results indicate that levels of atmospheric carbon dioxide must have been much higher in the past when the Sun was less powerful. Extra atmospheric carbon dioxide would have created a greater greenhouse effect and warmer temperatures. These results also indicate that some mechanism must have removed carbon dioxide from the atmosphere as the Sun's output of radiation increased over the Earth's geologic history. This mechanism is the conversion of atmospheric carbon dioxide into

fossilized organic matter (natural gas, oil, coal, limestone, and peat). In other words, Gaia!

This theory is important to Physical Geography and other Earth Sciences for the following reasons:

- The Gaia theory suggests that the abiotic and biotic environment is made up of many complex interrelationships;
- Many of these complex interrelationships are quite delicate and may be altered by human activity to a breaking point; and
- The theory suggests that humans must learn to respect Gaia by reducing their intentional modification of the Earth's abiotic and biotic components.

ENERGY AND MATTER

Energy is defined simply by scientists as the capacity for doing work. Matter is the material (atoms and molecules) that constructs things on the Earth and in the Universe. Albert Einstein suggested early in this century that energy and matter are related to each other at the atomic level. Einstein theorized that it should be possible to convert matter into energy. From Einstein's theories, scientists were able to harness the energy of matter beginning in the 1940s through nuclear fission. The most spectacular example of this process is a nuclear explosion from an atomic bomb. A more peaceful example of our use of this fact of nature is the production of electricity from controlled fission reactions in nuclear reactors. Einstein also suggested that it should be possible to transform energy into matter.

Energy and matter are also associated to each other at much larger scales of nature. Later on in this chapter, we will examine how solar radiation provides the energy to create the matter that makes up organisms. Organisms then use some of this matter to power their metabolism.

Types of Energy

Energy comes in a variety of forms. The simplest definition of the types of energy suggests that two forms exist: kinetic energy and potential energy. Kinetic energy is the energy due to motion. A rock falling from a cliff, a bee in flight, wind blowing leaves of trees, and water following over a waterfall are all examples of kinetic energy. Potential energy is the energy stored by an object that can be potentially transformed into another form of energy. Water stored behind a dam, the chemical energy of the food we consume, and the gasoline that we putting in our cars are all examples of potential energy. Conversion of this energy occurs when the energy in food is used by an organism to energize its metabolism, when the water in the dam flows through turbines to produce electricity from motion, and when the gasoline is used in a engine to produce motion from combustion.

Some other forms of energy include heat, electricity, sound, energy of chemical reactions, magnetic attraction, energy of atomic reactions, and light. Definitions for a few of these types of energy are as follows:

- Radiation - is the emission of energy from a material object in the form of electromagnetic waves and photons.
- Chemical Energy - is the energy produced or consumed in chemical reactions.
- Atomic Energy - is the energy released from an atomic nucleus because of a change in its subatomic mass.
- Electrical Energy - is the energy produced from the force between two objects having the physical property of electrical charge.
- Heat Energy - is a form of energy created by the combined internal motion of atoms in a substance.

On Earth, there are fundamentally only three ways in which energy can be transferred from one place to another: conduction, convection, and radiation. Conduction involves the adjacent transfer of heat energy from one atom to another through the mass of a gas, liquid, or solid. Condution results in the continuous flow of heat energy along a temperature gradient from areas of higher to lower temperature. Convection involves the transfer of heat energy by way of mass movements of a substance in gas or liquid form in a vertical direction (horizontal transfer is called advection). Convection is often seen as rising masses of gas or liquid called convection currents.

It is important to note that energy transfer by way of conduction and convection depends on the presence of matter. These forms of energy transfer do not operate in the vacuum of space. Radiation is the only means of energy transfer that can occur across outer space. The transfer of radiation produced at the Sun's surface through space supplies the Earth with most of its energy.

MATTER: ELEMENTS AND COMPOUNDS

Matter is the material that makes up things in the Universe. All matter on the Earth is constructed of elements. Chemists have described approximately 115 different elements. Each of these elements have distinct chemical characteristics. Table below lists some of the chemical characteristics for 48 common elements found in the Earth's continental crust.

The smallest particle that exhibits the unique chemical characteristics of an element is known as an atom. Atoms are composed of yet smaller particles known as protons, neutrons, and electrons. A proton is a subatomic particle that has significant mass and contributes a single positive electrical charge to an atom. Neutrons also have significant mass but no electrical charge. Electrons are extremely light subatomic particles having a mass that is 1/1840 of a proton. Each electron also has a negative electrical charge. Protons and neutrons make up the nucleus of an atom. As a result, most of an atom's mass is concentrated in the nucleus. Because protons are positively charged the nucleus has a positive charge equal to the number of these subatomic particles.

Electrons are found orbiting outside the nucleus at various distances based on their energy level. The area occupied by the electrons has a negative charge equal to the number of these subatomic particles.

If an atom has an equal number of electrons and protons its net electrical charge is zero. If there are more electrons than protons the charge of the atom is negative. Likewise, if there are less electrons than protons the charge of the atom is positive. In both cases, the exact charge is determined by subtracting protons from electrons. As a result, 4 protons minus 6 electrons give an atomic charge of -2.

The number of protons found in the nuclei of the different types of elements is unique and is referred to as the atomic number.

All atoms of a specific element have the same number of protons in their nuclei. Atomic mass number is an atom's total number of neutrons and protons. Many elements have unequal numbers of neutrons and protons in their nucleus. An element's atomic weight refers to the total weight of neutrons, protons, and electrons.

For example, the atomic weight of aluminum is 26.98. Atomic number describes the number of protons found in an atom. For example, silver has an atomic number of 47 or 47 protons in its atom. Some elements can have variants containing different numbers of neutrons but similar numbers of protons.

We call these variants isotopes. Carbon has two isotopes. Its most common form is carbon-12 which has 6 protons plus 6 neutrons.

About 99% of the carbon on our planet is of this type. The isotope carbon-13 has 6 protons plus 7 neutrons. Carbon-14 is the rarest isotope of carbon containing 8 neutrons. Some isotopes are unstable and their nucleus tends to lose subatomic particles forming an element with a lower atomic mass. This process is known as radioactive decay.

Elements can be classified as being either metals, nonmetals, or metalloids. Metals are elements that usually conduct heat and electricity and are shiny. Nonmetals do not conduct electricity that well and are normally not shiny. Metalloids have characteristics that are in between metals and nonmetals.

Elements with a net positive or negative charge are called ions. Chemists indicate the number of positive or negative charges on an ion using a superscript after the element's symbol. For example, calcium has two positive charges and is written as Ca^{2+}. Some common negatively charged ions include nitrate (NO^{3-}), sulfate (SO_4^{2-}), and phosphate (PO_4^{3-}).

Positive and negative ions are electrically attracted to each other. This mutual attraction allows for the bonding of atoms to occur forming structures of matter that are larger than just one atom. When similar atoms bond together they construct molecules. Atoms of different elements joined together form compounds. Sodium chloride (or table salt), is an ionic compound consisting of sodium (Na^+) and chloride (Cl^-). In nature, it forms as a three-dimensional array of oppositely charged ions. Many of the Earth's substances have a molecular structure similar to sodium chloride.

Inorganic vs. Organic

Compounds and molecules constructed in living tissues are commonly called organic. Forms of matter not formed by living things are termed inorganic. Organisms like autotrophs usually create organic matter by consuming inorganic molecules and compounds from the lithosphere, hydrosphere, and atmosphere. An example of an autotroph is any photosynthesizing plant. Heterotrophs consume and assimilate other living things to create their organic matter. Herbivores and carnivores are examples of heterotrophs.

There are four general categories of organic compounds: lipids, carbohydrates, proteins, and nucleic acids.

- Lipids - are composed of carbon atoms that have two hydrogen atoms attached. Lipids are commonly known as fats and oils, and belong to the family of molecules known as hydrocarbons.
- Carbohydrates - are composed of carbon, oxygen, and hydrogen atoms. Some examples are sugars, starch, and cellulose.
- Proteins - are organic compounds that are made primarily of carbon, hydrogen, nitrogen, and some other minor elements that are arranged into 20 different compounds known as amino acids.
- Nucleic Acids - are composed primarily of different combinations of carbon, hydrogen, nitrogen, oxygen, and phosphorus. They are very complex compounds being created by the atomic linking of thousands of individual atoms. DNA or deoxyribonucleic acid, the genetic blueprint of life, is an example of a nucleic acid.

Cells

All organisms are composed of one or more of cells. Cells are the smallest self-functioning unit found in living organisms. Cells are also where the processes of metabolism and heredity occur in an organism. Cells arise by the cellular division of a previously existing cell. Biologists have differentiated two basic types of cells in organisms. Bacteria, archaea, and cyanobacteria have cells that are quite uncomplicated in terms of structure and function. Quite simply, they lack internal organization. These cells are commonly known as prokaryotes.

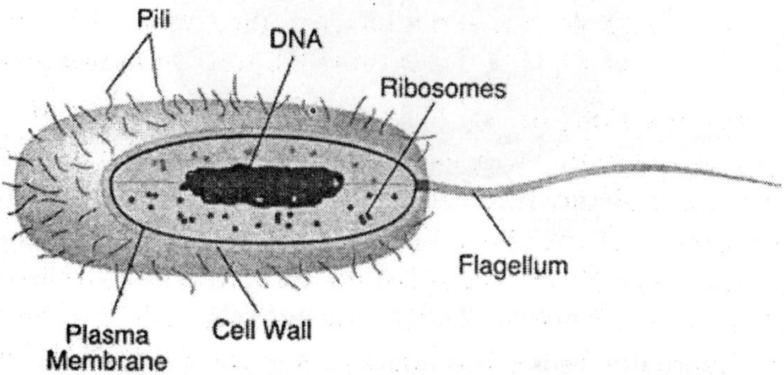

Figure: Typical prokaryote cell. These cells are about 1 to 10 micrometers in size. The cell is encased by rigid cell wall and a plasma membrane. Within the cell, the two most obvious structures are ribosomes and DNA. The DNA is not bounded by a membrane. Many prokaryote cells also have a flagellum which is used for movement.

The cells of plants and animals are more complex than those of bacteria, archaea, and cyanobacteria. We call these type of cells eukaryotes. Eukaryotic cells have a membrane enclosed nucleus which contains the organism's DNA. Plant and animal cells also contain a variety of membrane-bound structures known as organelles.

Eukaryotic cells also show great variation in terms of size. White blood cells of mammals are some of the smallest eukaryotic cells with a diameter between 3 to 4 micrometers. Ostrich ovum are very large cells with a diameter of about 100 micrometers.

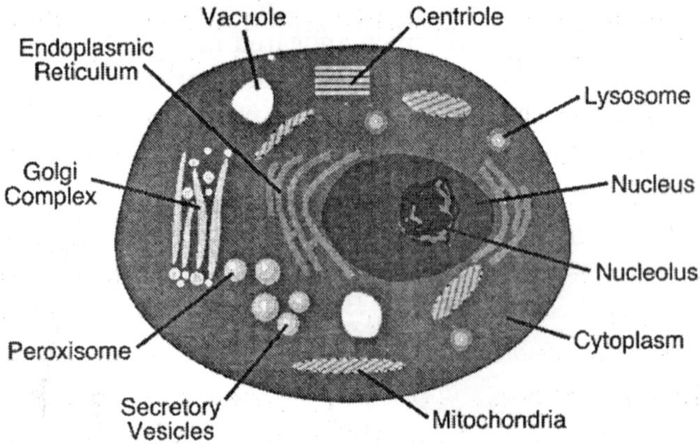

Figure: Typical animal cell.

Internally, cells contain specialized structures known as organelles that carry out distinct cellular functions. Within these structures enzymes, a type of protein, are used to facilitate and regulate various chemical reactions. Table below describes the function of a variety of cell structures including many organelles.

Cells can also be classified according to how they obtain their energy. Some cells have the ability to use light or chemical energy found in the outside environment to manufacture their own sugars, fats, and proteins. We call these types of cells autotrophs.

All 400,000 species of plants and a few species of bacteria use sunlight and the process of photosynthesis to obtain their energy. Some bacteria breakdown molecules found in the environment to release chemical energy to sustain their life. Organisms can also obtain their energy by consuming other organisms. These organisms are called heterotrophs. Heterotrophs include most types of bacteria and all of the animal and fungi species.

Some organisms consist of just one cell (bacteria, algae, and protozoa). However, most organisms are multi-cellular. Within multi-cellular organisms, groups of cells can become specialized to carry out a certain function. We call these functional groups of cells a tissue. An organ is a structure composed of several different types of tissues. Organs also have a specific structure and a particular function.

Measurement of Energy

In the previous section, we developed the concept of energy. We now must be able to measure and quantify it, using a standard set of units. Worldwide, two systems of units of measurement are commoly used today: the Metric System (Systeme International) and the British System. The units of energy described in these systems are derived from a technical definition of energy used by physicists. This definition suggests that energy can be represented by the following simple equation:

Work = Force × Distance

Similar to the definition given in the previous topic, physicists view energy as the ability to do work. However, they define work as a force applied to some form of matter (object) multiplied by the distance that this object travels. Physicists commonly describe force with a unit of measurement known as a newton (after Sir Isaac Newton). A newton is equal to the force needed to accelerate (move) a mass weighting one kilogram one meter in one second in a vacuum with no friction. The work or energy required to move an object with the force of one newton over a distance of one meter is called a joule. Some other definitions for the energy measurement units that you may come across in this textbook are as follows:

Calorie - equals the amount of heat required to raise 1 gram of pure water from 14.5 to 15.5° Celsius at standard atmospheric pressure. 1 calorie is equal to 4.1855 joules. The abreviation for calorie is cal. A kilocalorie, abbreviated kcal, is equal to a 1000 calories. 1 kilocalorie is equal to 4185 joules.

Btu - also called British thermal unit is the amount of energy required to raise the temperature of one pound of water one degree Fahrenheit.

Watt (W/m^2 or Wm^{-2}) - a metric unit of measurement of the intensity of radiation in watts over a square meter surface. One watt is equal to one joule of work per second. A kilowatt (kW) is the same as 1000 watts.

ENERGY, TEMPERATURE, AND HEAT

So far we have learned that energy can take on many forms. One important form of energy, relative to life on Earth, is kinetic energy. Simply defined, kinetic energy is the energy of motion. The amount of kinetic energy that a body possesses is dependent on the speed of its motion and its mass. At the atomic scale, the kinetic energy of atoms and molecules is sometimes referred to as heat energy.

Kinetic energy is also related to the concept of temperature. Temperature is defined as the measure of the average speed of atoms and molecules. The higher the temperature, the faster these particles of matter move. At a temperature of -273.15° Celsius (absolute zero) all atomic motion stops.

Heat is often defined as energy in the process of being transferred from one object to another because of difference in temperature between them. Heat is commonly transferred around our planet by the processes of conduction, convection, advection, and radiation.

Some other important definitions related to energy, temperature, and heat are:

- *Heat Capacity* - is the amount of heat energy absorbed by a substance associated to its corresponding temperature increase.
- *Specific Heat* - is equivalent to the heat capacity of a unit mass of a substance or the heat needed to raise the temperature of one gram (g) of a substance one degree Celsius. Water requires about 4 to 5 times more heat energy to raise its temperature when compared to an equal mass of most types of solid matter. This explains

why water bodies heat more slowly than adjacent land surfaces.

- *Sensible Heat* - is heat that we can sense. A thermometer can be used to measure this form of heat. Several different scales of measurement exist for measuring sensible heat. The most common are: Celsius scale, Fahrenheit scale, and the Kelvin scale.
- *Latent Heat* - is the energy needed to change a substance to a higher state of matter. This same energy is released from the substance when the change of state (or phase) is reversed. The diagram below describes the various exchanges of heat involved with 1 gram of water.

Energy and Life

The capture and use of energy in living systems is dominated by two processes: photosynthesis and respiration. Through these two processes living organisms are able to capture and use all of the energy they require for their activities.

Photosynthesis

Plants can capture the electromagnetic energy from the Sun by a chemical process called photosynthesis. This chemical reaction can be described by the following simple equation:

$$6CO_2 + 6H_2O + \text{light energy} >>> C_6H_{12}O_6 + 6O_2$$

The product of photosynthesis is the carbohydrate glucose and oxygen which is released into the atmosphere. All of the sugar glucose is produced in the specialized photosynthetic cells of plants and some other organisms. Glucose is produced by chemically combining carbon dioxide and water with sunlight.

This chemical reaction is catalyzed by chlorophyll acting in concert with other pigment, lipid, sugars, protein, and nucleic acid molecules. Sugars created in photosynthesis can be later converted by the plant to starch for storage, or it can be combined with other sugar molecules to form specialized carbohydrates such as cellulose, or it can be combined with other nutrients such as nitrogen, phosphorus, and sulfur, to build complex molecules such as proteins and nucleic acids.

Because all the energy fixed by a plant is converted into sugar, it is theoretically possible to determine a plant's energy uptake by measuring the amount of sugar produced. This quantity is called gross primary productivity. Measurements of the buildup of sugar in the plant reflect gross primary productivity less respiration, or net primary productivity.

In general, animals cannot produce their own energy via photosynthesis. Instead, they capture their energy by the consumption and assimilation of the biomass of plants or other animals. Thus, animals get the energy they need for maintenance of their bodies tissues, growth, and reproduction indirectly from photosynthetic organisms.

Respiration

The oxidation of sugar by organisms is called respiration. This process occurs in both plants and animals. In most organisms, respiration releases the energy required for all metabolic processes. This chemical reaction can be described by the following simple equation:

$$C_6H_{12}O_6 + 6O_2 >>> 6CO_2 + 6H_2O + \text{released energy}$$

One of the products of respiration is energy, which is released via the chemical decomposition of glucose. Other products of this chemical reaction are carbon dioxide (CO_2) and water (H_2O).

Laws of Thermodynamics

The field of thermodynamics studies the behavior of energy flow in natural systems. From this study, a number of physical laws have been established. The laws of thermodynamics describe some of the fundamental truths of thermodynamics observed in our Universe. Understanding these laws is important to students of Physical Geography because many of the processes studied involve the flow of energy.

First Law of Thermodynamics

The first law of thermodynamics is often called the Law of Conservation of Energy. This law suggests that energy can be transferred from one system to another in many forms. Also, it can not be created or destroyed. Thus, the total amount of energy available in the Universe is constant. Einstein's famous equation (written below) describes the relationship between energy and matter:

$$E = mc^2$$

In the equation above, energy (E) is equal to matter (m) times the square of a constant (c). Einstein suggested that energy and matter are interchangeable. His equation also suggests that the quantity of energy and matter in the Universe is fixed.

Second Law of Thermodynamics

Heat cannot be transfer from a colder to a hotter body. As a result of this fact of thermodynamics, natural processes that involve energy transfer must have one direction, and all natural processes are irreversible.

This law also predicts that the entropy of an isolated system always increases with time. Entropy is the measure of the disorder or randomness of energy and matter in a system. Because of the second law of thermodynamics both energy and matter in the Universe are becoming less useful as time goes on. Perfect order in the Universe occurred the instance after the Big Bang when energy and matter and all of the forces of the Universe were unified.

Third Law of Thermodynamics

The third law of thermodynamics states that if all the thermal motion of molecules (kinetic energy) could be removed, a state called absolute zero would occur. Absolute zero results in a temperature of 0 Kelvins or -273.15° Celsius.

Absolute Zero = 0 Kelvins = -273.15° Celsius

The Universe will attain absolute zero when all energy and matter is randomly distributed across space. The current temperature of empty space in the Universe is about 2.7 Kelvins.

The Nature of Radiation

All objects above the temperature of absolute zero (-273.15° Celsius) radiate energy to their surrounding environment. This energy, or radiation, is emitted as electromagnetic waves that travel at the speed of light. Many different types of radiation have been identified. Each of these types is defined by its wavelength. The wavelength of electromagnetic radiation can vary from being infinitely short to infinitely long.

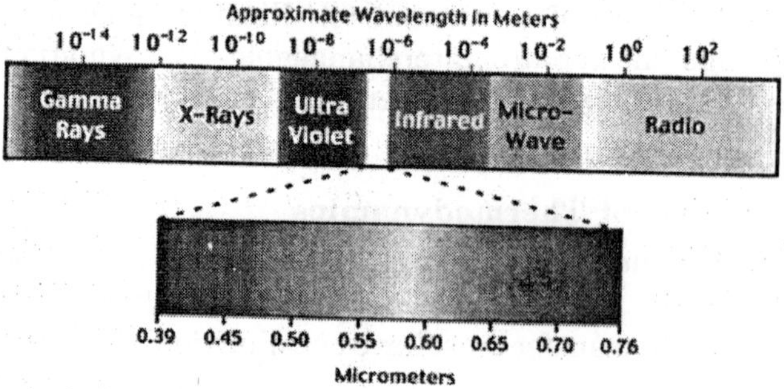

Figure: Some of the various types of electromagnetic radiation as defined by wavelength. Visible light has a spectrum that ranges from 0.40 to 0.71 micrometers (μm).

Visible light is a form of electromagnetic radiation that can be perceived by our eyes. Light has a wavelength of between 0.40 to 0.71 micrometers (μm). Figure above illustrates that various spectral color bands that make up light.

$$E^* = \sigma T^4$$

where σ (sigma) = $5.67 \times 10^{-8}\ Wm^{-2}\ K^{-4}$

and T is the temperature in Kelvin

According to the Stephan-Boltzmann equation, a small increase in the temperature of a radiating body results in a large amount of additional radiation being emitted.

The Sun emits only a portion (44 %) of its radiation in zone. Solar radiation spans a spectrum from approximately 0.1 to 4.0 micrometers. The band from 0.1 to 0.4 micrometers is called ultraviolet radiation. About 7% of the Sun's emission is in this wavelength band. About 48% of the Sun's radiation falls in the region between 0.71 to 4.0 micrometers. This band is called the near (0.71 to 1.5 micrometers) and far infrared (1.5 to 4.0 micrometers).

The amount of electromagnetic radiation emitted by a body is directly related to its temperature.

If the body is a perfect emitter (black body), the amount of radiation given off is proportional to the 4th power of its temperature as measured in Kelvin units. This natural phenomenon is described by the Stephan-Boltzmann Law. The following simple equation describes this law mathematically:

In general, good emitters of radiation are also good absorbers of radiation at specific wavelength bands.

This is especially true of gases and is responsible for the Earth's greenhouse effect. Likewise, weak emitters of radiation are also weak absorbers of radiation at specific wavelength bands. This fact is referred to as Kirchhoff's Law.

Some objects in nature have almost completely perfect abilities to absorb and emit radiation. We call these objects black bodies. The radiation characteristics of the Sun and the Earth are very close to being black bodies.

The wavelength of maximum emission of any body is inversely proportional to its absolute temperature. Thus, the higher the temperature, the shorter the wavelength of maximum emission. This phenomenon is often called Wien's Law. The following equation describes this law:

$$\lambda_{max} = C/T$$

where C is a constant equal to 2897
and T is the temperature in Kelvin

Wien's law suggests that as the temperature of a body increases, the wavelength of maximum emission becomes smaller.

According to the above equation the wavelength of maximum emission for the Sun (5800 Kelvins) is about 0.5

micrometers, while the wavelength of maximum emission for the Earth (288 Kelvins) is approximately 10.0 micrometers.

A graph that describes the quantity of radiation that is emitted from a body at particular wavelengths is commonly called a spectrum. The following two graphs describe the spectrums for the Sun and Earth.

The Solar Source of the Earth's Energy

Almost all of the energy that drives the various systems (climate systems, ecosystems, hydrologic systems, etc.) found on the Earth originates from the Sun.

Solar energy is created at the core of the Sun when hydrogen atoms are fused into helium by nuclear fusion.

For each second of this nuclear process, 700 million tons of hydrogen are converted into 695 million tons of helium. The remaining 5 million tons are turned into electromagnetic energy that radiates from the Sun's surface out into space.

The radiative surface of the Sun, or photosphere, has an average temperature of about 5800 Kelvins. Most of the electromagnetic radiation emitted from the Sun's surface lies in the visible band centered at 0.5 μm. The total quantity of energy emitted from the Sun's surface is approximately 63,000,000 Watts per square meter (W/m^2 or Wm^{-2}).

The energy emitted by the Sun passes through space until it is intercepted by planets and other celestial objects. The intensity of solar radiation striking these objects is determined by a physical law known as the Inverse Square Law.

This law merely states that the intensity of the radiation emitted from the Sun varies with the squared distance from the source. As a result of this law, if the intensity of radiation at a given distance is one unit, at twice the distance the intensity will become only one-quarter. At three times the distance, the intensity will become only one-ninth of its original intensity at a distance of one unit, and so on.

Given the amount of energy radiated by the Sun and the average Earth-Sun distance of 149.5 million kilometers, the amount of radiation intercepted by the outer limits of the atmosphere can be calculated to be around 1370 W/m^2. For general purposes, the energy output of the Sun can be considered constant. This of course is not entirely true. Scientists have shown that the output of the Sun is temporally variable. Some researchers have also suggested that the increase in the average global temperature over the last century may have been solar in origin. This statement, however, is difficult to prove because accurate data on solar output of radiation only goes back to about 1978.

EARTH-SUN GEOMETRY

Earth Rotation and Revolution

The term Earth rotation refers to the spinning of our planet on its axis. Because of rotation, the Earth's surface moves at the equator at a speed of about 467 m per second or slightly over 1675 km per hour. If you could look down at the Earth's North Pole from space you would notice that the direction of rotation is counter-clockwise.

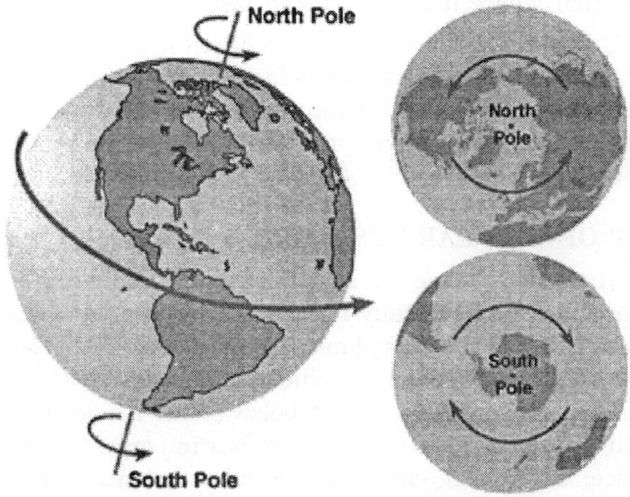

Figure: The movement of the Earth about its axis is known as rotation. The direction of this movement varies with the viewer's position. From the North Pole the rotation appears to move in a counter-clockwise fashion. Looking down at the South Pole the Earth's rotation appears clockwise.

The opposite is true if the Earth is viewed from the South Pole. One rotation takes exactly twenty-four hours and is called a mean solar day. The Earth's rotation is responsible for the daily cycles of day and night.

At any one moment in time, one half of the Earth is in sunlight, while the other half is in darkness. The edge dividing the daylight from night is called the circle of illumination. The Earth's rotation also creates the apparent movement of the Sun across the horizon.

The orbit of the Earth around the Sun is called an Earth revolution. This celestial motion takes 365.26 days to complete one cycle. Further, the Earth's orbit around the Sun is not circular, but oval or elliptical. An elliptical orbit causes the Earth's distance from the Sun to vary over a year. Yet, this phenomenon is not responsible for the Earth's seasons! This variation in the distance from the Sun causes the amount of solar radiation received by the Earth to annually vary by about 6%. Figure below illustrates the positions in the Earth's revolution where it is closest and farthest from the Sun.

On January 3, perihelion, the Earth is closest to the Sun (147.3 million km). The Earth is farthest from the Sun on July 4, or aphelion (152.1 million km). The average distance of the Earth from the Sun over a one-year period is about 149.6 million km.

TILT OF THE EARTH'S AXIS

The ecliptic plane can be defined as a two-dimensional flat surface that geometrically intersects the Earth's orbital path around the Sun. On this plane, the Earth's axis is not at right angles to this surface, but inclined at an angle of about 23.5° from the perpendicular. Figure below shows a side view of the Earth in its orbit about the Sun on four important dates: June solstice, September equinox, December solstice, and March equinox. Note that the angle of the Earth's axis in relation to the ecliptic plane and the North Star on these four dates remains unchanged. Yet, the relative position of the Earth's axis to the Sun does change during this cycle. This circumstance is responsible for the annual changes in the height of the Sun above the horizon. It also causes the seasons, by controlling the intensity and duration of sunlight received by locations on the Earth. Figure below shows an overhead view of this same phenomenon. In this view, we can see how the circle of illumination changes its position on the Earth's surface. During the two equinoxes, the circle of illumination cuts through the North Pole and the South Pole. On the June solstice, the circle of illumination is tangent to the Arctic Circle (66.5° N) and the region above this latitude receives 24 hours of daylight. The Arctic Circle is in 24 hours of darkness during the December solstice.

On June 21 or 22 the Earth is positioned in its orbit so that the North Pole is leaning 23.5° toward the Sun. During the June solstice (also called the summer solstice in the Northern Hemisphere), all locations north of the equator have day lengths greater than twelve hours, while all locations south of the equator have day lengths less than twelve hours. On December 21 or 22 the Earth is positioned so that the South Pole is leaning 23.5 degrees toward the Sun. During the December solstice (also called the winter solstice in the Northern Hemisphere), all locations north of the equator have day lengths less than twelve hours, while all locations south of the equator have day lengths exceeding twelve hours. On September 22 or 23, also called the autumnal equinox in the Northern Hemisphere, neither pole is tilted toward or away from the Sun. In the Northern Hemisphere, March 20 or 21 marks the arrival of the vernal equinox or spring when once again the poles are not tilted toward or away from the Sun. Day lengths on both of these days, regardless of latitude, are exactly 12 hours.

Axis Tilt and Solar Altitude

The annual change in the relative position of the Earth's axis in relationship to the Sun causes the height of the Sun or solar altitude to vary in our skies. Solar altitude is normally measured from either the southern or northern point along the horizon and begins at zero degrees. Maximum solar altitude occurs when the Sun is directly overhead and has a value of 90°.

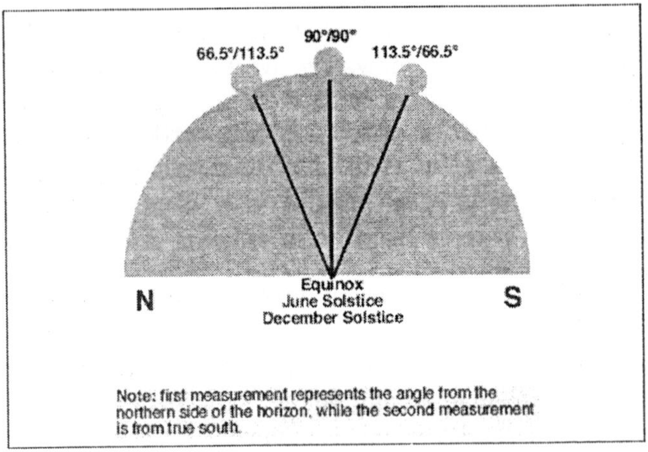

Figure: Variations in solar altitude at solar noon for the equator during the June solstice, equinox, and December solstice.

The total variation in maximum solar altitude for any location on the Earth over a one-year period is 47° (Earth's tilt 23.5° × 2 = 47°).

This variation is due to the annual changes in the relative position of the Earth to the Sun. At 50 degrees North, maximum solar altitude varies from 63.5 degrees on the June solstice to 16.5 degrees on the December solstice.

Maximum solar height at the equator goes from 66.5 degrees above the northern end of the horizon during the June solstice, to directly overhead on the September equinox, and then down to 66.5 degrees above the southern end of the horizon during the December solstice.

The location on the Earth where the Sun is directly overhead at solar noon is known as the subsolar point.

The subsolar point occurs on the equator during the two equinoxes. On these dates, the equator is lined up with the ecliptic plane and the poles are in line with the circle of illumination.

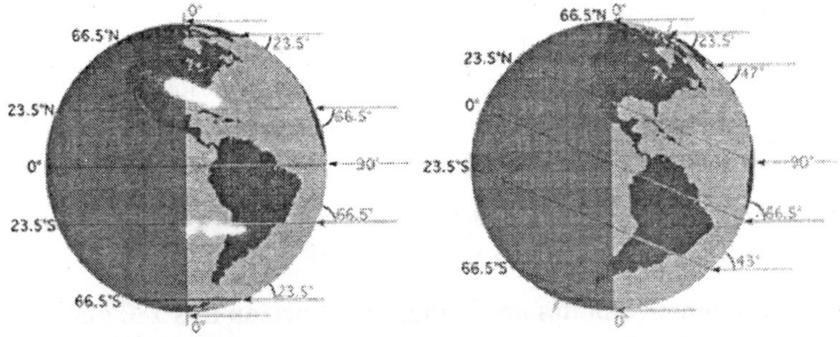

Figure: Relationship of maximum Sun height to latitude for the equinox (left) and June solstice (right). The red values on the right of the globes are maximum solar altitudes at solar noon. Black numbers on the left indicate the location of the Equator, Tropic of Cancer (23.5 degrees N), Tropic of Capricorn (23.5 degrees S), Arctic Circle (66.5 degrees N), and the Antarctic Circle (66.5 degrees S). The location of the North and South Poles are also identified. During the equinox, the equator is the location on the Earth with a Sun angle of 90 degrees for solar noon. Note how maximum Sun height declines with latitude as you move away from the Equator. For each degree of latitude traveled maximum Sun height decreases by the same amount. At equinox, you can also calculate the noon angle by subtracting the location's latitude from 90. During the summer solstice, the Sun is now directly overhead at the Tropic of Cancer. All locations above this location have maximum Sun heights that are 23.5 degrees higher from the equinox situation. Places above the Arctic Circle are in 24 hours of daylight. Below the Tropic of Cancer the noon angle of the Sun drops one degree in height for each degree of latitude traveled. At the Antarctic Circle, maximum Sun height becomes 0 degrees and locations south of this point on the Earth are in 24 hours of darkness.

During the summer solstice, the subsolar point moves to the Tropic of Cancer (23.5° N) because at this time the North Pole is inclined 23.5° toward the Sun.

Figure below shows how the subsolar point gradually changes from one day to the next over a period of one-year. Note that on this graph, the subsolar point is located at the Tropic of Capricorn (23.5° S) during the December solstice when the South Pole is angled 23.5° toward the Sun.

The following table describes the changes in solar altitude at solar noon for the two solstices and equinoxes.

All measurements are in degrees (horizon has 180 degrees from True North to True South) and are measured from either True North or True South (whatever is closer). The following links show graphical illustrations of the annual movements of the Sun in our skies for selected latitudes. In these illustrations, solar angles are measured from both True North and True South for solar noon.

| 90 N | 66.5 N | 50 N | 23.5 N | Equator (0) | 23.5 S | 50 S | 66.5 S | 90 S |

Finally, the altitude of the Sun at solar noon can also be calculated with the following simple equation:

Altitude A = 90 - Latitude L +/- Declination D

In this equation, L is the latitude of the location in degrees and D is the declination. The equation is simplified to A = 90 - L if Sun angle determinations are being made for either equinox date. If the Sun angle determination is for a solstice date, declination (D) is added to latitude (L) if the location is experiencing summer (northern latitudes = June solstice; southern latitudes = December solstice) and subtracted from latitude (L) if the location is experiencing winter (northern latitudes = December solstice; southern latitudes = June solstice).

All answers from this equation are given relative to True North for southern latitudes and True South for northern latitudes. For our purposes only the declinations of the two solstices and two equinoxes are important. These values are: June solstice D=23.5, December solstice D=23.5, March equinox D=0, and Septemeber equinox D=0.

When using the above equation in tropical latitudes, Sun altitude values greater than 90 degrees may occur for some calculations. When this occurs, the noonday Sun is actually behind you when looking towards the equator. Under these circumstances, Sun altitude should be recalculated as follows:

Altitude A = 90 - (originally calculated Altitude A - 90)

Earth-Sun Relationships and Insolation

In the previous topic, we learned that the Earth's seasons are controlled by changes in the duration and intensity of solar radiation or insolation. Both of these factors are in turn governed by the annual change in the position of the Earth's axis relative to the Sun.

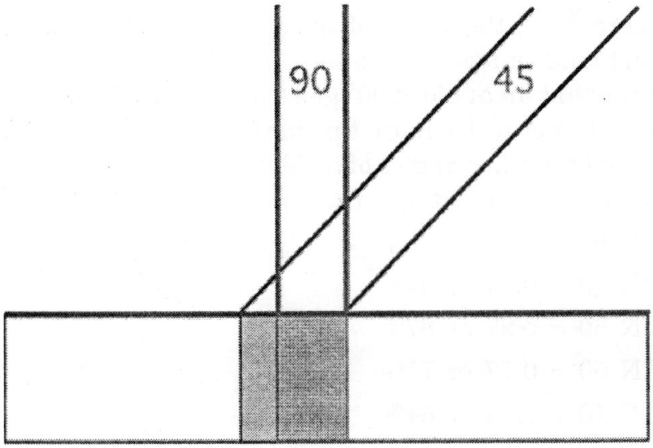

Figure: Effect of angle on the area that intercepts an incoming beam of radiation.

Yearly changes in the position of the Earth's axis cause the location of the Sun to wander 47° across our skies. Changes in the location of the Sun have a direct effect on the intensity of solar radiation.

The intensity of solar radiation is largely a function of the angle of incidence, the angle at which the Sun's rays strike the Earth's surface. If the Sun is positioned directly overhead or 90° from the horizon, the incoming insolation strikes the surface of the Earth at right angles and is most intense. If the Sun is 45° above the horizon, the incoming insolation strikes the Earth's surface at an angle. This causes the rays to be spread out over a larger surface area reducing the intensity of the radiation. Figure below models the effect of changing the angle of incidence from 90 to 45°. As illustrated, the lower Sun angle (45°) causes the radiation to be received over a much larger surface area. This surface area is approximately 40% greater than the area covered by an angle of 90°. The lower angle also reduces the intensity of the incoming rays by 30%.

We can also model the effect the angle of incidence has on insolation intensity with the following simple equation:

Intensity = SIN (A)

where, A is the angle of incidence and SIN is the sine function found on most calculators. Using this equation we can determine that an angle of 90° gives us a value of 1.00 or 100% (1.00 x 100). Let us compare this maximum value with values determined for other angles of incidence. Note the answers are expressed as a percentage of the potential maximum value.

SIN 80 = 0.98 or 98%

SIN 70 = 0.94 or 94%

SIN 60 = 0.87 or 87%

SIN 50 = 0.77 or 77%

SIN 40 = 0.64 or 64%

SIN 30 = 0.50 or 50%

SIN 20 = 0.34 or 34%

SIN 10 = 0.17 or 17%

SIN 0 = 0.00 or 0%The yearly changes in the position of the Earth's axis relative to the plane of the ecliptic also causes seasonal variations in day length to all locations outside of the equator. Longest days occur during the June solstice for locations north of the equator and on the December solstice for locations in the Southern Hemisphere.

The equator experiences equal day and night on every day of the year. Day and night is also of equal length for all Earth locations on the September and March equinoxes. Figure below describes the change in the length of day for locations at the equator, 10, 30, 50, 60, and 70 degrees North over a one-year period. The illustration suggests that days are longer than nights in the Northern Hemisphere from the March equinox to the September equinox. Between the September to March equinox days are shorter than nights in the Northern Hemisphere. The opposite is true in the Southern Hemisphere. The graph also shows that the seasonal (winter to summer) variation in day length increases with increasing latitude.

6

THEORY, REALITY AND PRACTICE OF GEOGRAPHY

Defining theory precisely is highly problematic. Some authors provide rigid and highly formalized definitions of theory, whilst others question the value of the term at all. Theory can be viewed as a framework of ideas that guide what we think reality is and how to go about studying it. Abstract ideas such as force and resistance can be thought of as theoretical constructs. These two ideas can be linked together to suggest what should happen in reality. For example, if force is greater than resistance then change should occur. The two ideas form irreducible elements of the theory. They are the axiomatic elements. Their existence is not questioned; it is taken as given within the theory. Likewise, their interactions, usually based upon some set of physical principles or mechanisms, are the axiomatic principles of the theory.

The ideas are, however, very abstract and need to be translated or superimposed upon parts of reality. Theory needs to be linked to reality. This is usually achieved by bridging principles, rules that define how abstract ideas can be interpreted as real entities capable of identification and testing. Within science, these bridging principles are essential as the means by which theories can be tested and, in a pure Popperian sense, rejected or accepted.

It is interesting to note that Haines-Young and Petch (1986) distinguish between theories and myths. Within myths they include Schumm and Lichty's (1965) paper on time, space and causality, Wolman and Miller's (1960) paper on magnitude and frequency, as well as Hack's (1960) paper on dynamic equilibrium. For many geomorphologists in particular these papers would seem to be central to any understanding of the physical environment. Haines-Young and Petch, however, make the point that all these papers express what they term as 'truisms'.

For Schumm and Lichty, for example, they state that the truism is 'that theories explain variables and that only certain theories have been developed' (p. 139), a problem at the time of the paper, but, in their eyes, a trivial issue in retrospect. Wolman and Miller (1960) are merely stating the truism that whatever landscape feature is observed, it must have been formed by processes that had some magnitude and frequency (p. 139).

These papers, as well as Hack's (1960) outline of dynamic equilibrium, state ideas that cannot be disproved. They can be linked to reality in a multitude of ways excluding nothing, and can never be falsified because of this. Inability to focus or narrow down a set of ideas to form a distinct and singular set of bridging principles means that there is always scope for reinterpretation of the ideas in a favourable light; that is, the ideas are unfalsifiable. Myths are not useless in physical geography, however. Haines-Young and Petch point out that although these ideas are not theories, as defined within critical rationalism, and so are not sources of any understanding about reality, such myths form the basis of a lot of testable theories in science.

They view it as possible to derive testable and coherent theories from the myths. They see it as the task of a scientist to develop such links between myth and theory, to prompt questions based on these myths that are amenable to testing in reality.

Myths and Theories

The classic paper on 'Time, space and causality' by Schumm and Lichty (1965) is often viewed as a major turning point in thinking about geomorphology. Haines-Young and Petch (1986), however, suggest that the main argument of the paper amounts to a 'truism' or myth rather than a theory. The basic idea is that at different temporal scales, different variables will be independent, dependent or irrelevant for understanding the functioning of the landscape. The temporal scale of analysis you are interested in determines which of the variables you need to study. Undertaking a study of a catchment within graded time, for example, would require an analysis of relationship between dependent and independent variables. Variables such as time and initial relief are irrelevant for your study as these are unchanging over this time-scale.

Haines-Young and Petch (1986) characterize this point as merely the statement that geomorphologists have developed different theories to explain landform behaviour at different scales. The variables vary in their status as dependent, independent or irrelevant, because the theories used change. It is not the variables that alter their status, it is theories that alter to explain a different variable at a specific scale. Tinkler (1985) notes that the key point to Schumm and Lichty's analysis was to point out that theories actually relate to different things at different scales. Acceptance of one approach, the historical, is not incompatible with another, say that of equilibrium, as each deals with different things at different scales. Each is mediated by different theories that identify different objects of study.

The ideas of Schumm and Lichty are not falsifiable. They are only statements of how research into reality is structured within geomorphology. The idea that theories relate to different variables at different scales does not generate any hypotheses to assess. The idea could, however, provide a basis for directing thinking about reality. If theories relate to different variables at different scales, then an investigator is not bound by an overarching single scale-independent theory. What is an appropriate theory, an appropriate approach to study, could vary with the scale of object being studied.

Indeed, the objects themselves could change as you change scale. It is in this sense that the 'myth' of scale put forward by Schumm and Lichty is of use in geomorphology. Line is between the myth and the derived theory is unclear. There may be many ways of undertaking such a translation from myth to theory, and so the problem of an unfalsifiable base remains, despite the presence of a falsifiable product, a theory. Von Engelhardt and Zimmerman (1988) suggest that theories are hierarchically ordered systems of hypotheses interlocked by a network of deductive relationships.

This highlights the significance of hypotheses to one view of theory and possibly may give myths a role as a higher-level structure. They view theory, hypotheses and the empirical levels of reality as interlinked. Using the theory of plate tectonics they construct a hierarchical framework of theories, hypotheses and 'facts'.The highest level of the theory are the basic hypotheses from which the partial hypotheses are derived.

From the partial hypotheses are derived the inductive generalizations and empirical basis of the theory. Each level informs the others. Note, however, that the basic hypotheses are few but that the hypotheses derived from them are many. Likewise, the empirical 'facts' are more in number than the partial hypotheses that they inform.

The structure of an increasing number of derived statements is a common one. Both empirical and theoretical 'facts' (by which they mean theoretical concepts such as the atom) can both be used to construct theories. Within physical geography it can be assumed that most theories will contain both and be informed by the success or failure to explain or predict the behaviour of these entities. Von Engelhardt and Zimmerman draw little distinction between theory and hypotheses regarding the two as related to each other.

Theories only differ in being further up the hierarchy and being greater in the scope of their application, their significance as explanatory structures, and being more reliable in terms of consistency. According to von Engelhardt and Zimmerman (1988), a theory should be both internally consistent, in terms of its logic structure, and externally consistent, in terms of its consistency with the hypotheses derived from other theories. These other theories include the underlying physical and chemical theories that usually make up unquestioned components of theories in physical geography. They do recognize that both theory and hypothesis represent generalization, abstractions of reality. They are not mirror images of reality, they describe and explain only the empirical and hypothetical components of reality and so are only indirectly reflecting reality as it is.

Theories can be identified at a number of levels in any study. A theory about force and resistance within fluvial systems requires bridging principles to link its elements to measurable entities within the fluvial system. If force is equated with flow velocity then measurement of flow velocity becomes a surrogate for this part of the theory. But how can flow velocity be identified and measured? Flow velocity will vary across the river profile, so there need to be criteria for selecting precisely where flow is measured.

Watching a river you may notice that flow need not always be directly downstream. River flows swirl and are deflected, forming eddy currents of various sizes. Again agreement is required concerning the appropriate point or section of river to measure flow. The instruments used to measure flow can vary from an orange and a stop-watch along a measured reach to an acoustic or a laser Doppler device. How can we be sure that these instruments are measuring the same thing – river flow? Instruments such as the laser Doppler-flow monitor measure movement based on specific physical principles about how reality works. In other words, our theory of river flow as a force is being measured by an instrument that is itself designed and developed on the basis of a theory about reality. If the principles upon which the instrument is based are wrong then the measurement of flow is flawed.

All observations are theory-laden to some degree or other, as Rhoads and Thorn (1996b) noted. The question is, does this theory-ladenness matter to the final identification and testing of the theory at the level of interest to the physical geographer? Identifying that laser Doppler devices are dependent upon a theory about how reality works developed within physics is something inherent to all instruments of this type. The theories associated with the production of laser light, the interaction of the laser with the environment, are all assumed to be correct. Without this assumption there could be no assessment of theories developed at the level of the entities of interest to physical geography. We would instead be continually questioning and testing the physics and chemistry of our methods.

Bogen and Woodward (1988) provide an interesting spin on theory-ladenness, a viewpoint developed by Basu (2003) in relation to the work of Priestley and Lavoisier on oxygen. They suggest that, contrary to popular belief, theories do not make predictions or explanations about facts. Instead they believe that a distinction has to be made between data and phenomena. Data have the task of providing evidence for the existence of a phenomenon and are relatively easy to observe. Data, however, cannot be predicted or explained that well by theory – data just are. Phenomena, on the other hand, are detected through the use of data, but, in all but the most exceptional cases, are not observable. Data stand as potential indicators of a phenomenon, but not necessarily just one phenomenon. Data could be used to identify many different phenomena depending on what the data are thought to detect. Phenomena are viewed as being generated by establishing a particular experimental design.

The design is produced to identify and monitor the phenomena. There are a limited number of circumstances that will permit the phenomena to emerge from the experimental situation. It would be expected, however, that phenomena would emerge once the limited circumstances had been obtained. Data involve such disparate and numerous combinations of factors, often unique to the experimental situation, that no general theory can be constructed to explain or predict data.

Phenomena have stable and repeatable characteristics that can be detected, provided the appropriate circumstances are present, by a variety of different types of data. The different types of data need to be produced by the phenomena in sufficient and appropriate quantities to enable their detection and interpretation within the theory being assessed.

Data cannot, therefore, be produced by any means, they need to be produced within a context appropriate for the assessment of a phenomenon. This means that data production is informed by the theory under assessment and will be restricted in its location and timing of production as dictated by the phenomenon under study. If a phenomenon can only be isolated in the laboratory, then only data produced within the experimental confines of the laboratory will be considered as relevant. Production of such data may involve long processes of establishing adequate control and techniques for ensuring data quality, as well as alteration of the data to a form amenable to appropriate statistical analysis.

Physical Geography as Historical Science

There has been a tendency, both historically and currently, to try to set the working practices of physical geography as a separate type of scientific explanation: as a historical science. Some practitioners have assumed the inferior status of a historical science. Andersson (1996), for example, in relation to explanation in historical biogeography, stated that the explanatory structure of mathematics, physics and chemistry was clearly the most prestigious model of explanation with the greatest explanatory power, but one hardly applicable to historical biogeography.

Johnson (1933) identifies seven stages to investigation in historical sciences. Within each stage he believes it is vital to undertake detailed analysis. By analysis he means detailed and thoughtful testing of the basis of each investigation at each stage or 'tracing back each part to its source and testing its validity, for the purpose of clarifying and perfecting knowledge'. His preferred method for studying the environment is by the use of multiple working hypotheses, through which different ideas can be tested and analysed.

The basis for distinguishing the practice and explanations developed in physical geography – whether it is in the guise of ecology or geology – from those of 'traditional' or experimental science, lies in the subject matter of study and the explanatory framework sought. A historical science is not necessarily concerned with generalities, instead it is concerned with explaining particular events, both present and past: Similarly, in geology we are largely interested in historical 'individuals' (this outcrop, the Western Interior Seaway, the lifespan of a species) and their specific life history. It is possible to identify general laws in geology that have explanatory power . . . but the weight of our interest lies elsewhere. In explaining specific individuals and events reference may be made to general laws derived from 'hard' sciences, but they will be used as part of the logical arguments concerning the specific individuals or events. The 'laws' will not be the explanation of the specific, instead they will form part of the explanatory framework.

The historical scientist has observed phenomena as their focus of study, but without the luxury of observed causes (Cleland, 2001). In experimental sciences, causes are observed, often under controlled conditions, and phenomena can be observed as produced by them under the controlled conditions.

Historical science has to deal with what Cleland (2001) calls the asymmetry of localized events. An event such as an eruption produces a multitude of effects of which only a few need to be identified to infer that an eruption has occurred. A single effect cannot, however, be used to infer a cause. There are many possible and plausible causes that can be linked to a single effect. Historical sciences have developed strategies – that of multiple working hypotheses being one – with which to whittle down the number of cause-and-effect relationships. Historical science searches for the most plausible set of cause-and-effect links as the explanation for a phenomenon.

Historical science looks for what Cleland calls the 'smoking gun', the piece of evidence that unequivocally links a specific cause to a specific effect. Failing a singular, decisive piece of evidence, historical science instead focuses on identifying a subset of effects, or 'traces' as Cleland calls them, which can be viewed as evidence that only a specific cause could have produced them.

Tradition and experience will identify the small subset of traces or types of evidence that usually provide information for this judgement. Within environmental reconstruction, for example, pollen analysis or stratigraphy is probably the most widely adopted and versatile method (Lowe and Walker, 1997).

Pollen can be extracted from different sediments and used as evidence for a whole range of environmental conditions.

Causation within historical science is recognized as a complicated and potentially complex affair. Sets of traces, or evidence, are built up to construct a picture of the past and the manner in which past events caused effects in the past and resulted in the present state of affairs. It is often unclear where causation lies within this picture.

The bursting of a glacial dam might produce a catastrophic flood that alters the course of a river and is a major formative event in the development of a number of landforms. Identifying evidence of the dam-burst might be supposed to have identified the cause of the changing river course or of the slope profiles of the catchment. The dam-burst may have triggered the effects, but is it the cause? The flood of water would have had a lesser impact had the river channel been highly constrained by existing geology. Likewise, a large flat floodplain with few steep slopes would not be affected by the catastrophic flood in the same manner as a newly deglaciated landscape with unconsolidated moraines scattered across it.

In other words, the context within which the trigger event occurs is as important as the trigger itself in any explanation. Simpson (1963) noted that the immanent or ahistorical processes, that experimental science is good at identifying, need to be considered with the configurational aspects of a situation to understand or explain reality.

The configurational is the series of states that have uniquely occurred through the interaction of ahistorical processes with historical circumstance – the context. Without an understanding of both, causation may be mistakenly attributed to a single event. Of itself, the trigger is not sufficient to cause the effect; the trigger is only a cause because of the specific configuration of reality it occurs within.

It is worth noting that historical science also has a major problem with identifying the nature of the individuals it is trying to study (Hull, 1978, 1981). Individuals do not appear fully defined and fully formed for study.

A set of characteristics are needed to differentiate the individual from its context. Furthermore, the individual may alter in these characteristics over time, as its individuality is degraded and becomes increasingly undifferentiated from its context. For example, as a mass of material moves down a slope at what point is that mass no longer a part of the slope? When does it become an individual in its own right? Alternatively, as soil properties change, moving across a landscape, at what point does one soil type become another? Although there may be standardized classification criteria for identifying soil types, is a distinct value at which a soil is one thing and then another appropriate to understanding reality?

Summary

Physical geographers study a reality that they assume is external to themselves and that is capable of study. Although they may never have absolute knowledge of this external reality they can organize and use information to argue and come to an agreement about the nature of their representations of reality.

Logical positivism believes that it is possible to sense reality as it is and that it is possible to discover real laws about real entities through sensor information. Research is directed towards proving reality to be as we believe it to be. Critical rationalism takes a more critical view of our search for an understanding of reality. Critical rationalists view the testing and falsifying of ideas as the essential feature of scientific research. Knowledge is never certain or absolute, it is only the ideas which we cannot disprove that we accept as true for the time being. Within both philosophies deductive and inductive reasoning are used to reason about reality. Deduction is the only logically consistent method of argument, but unless it is connected to reality via empirical assessment it can produce invalid arguments.

Critical realism highlights the differentiated and stratified nature of reality, the asymmetry between the real and the actual. All we can observe is the actual, but this is a pale reflection of the structures and mechanisms that underlie reality. Assessment of ideas is more problematic in critical realism, but testing of hypotheses about reality is still a central means of understanding. Pragmatism emphasizes the importance of the mental image the researcher has of reality in understanding that reality. Any testing of ideas makes use of entities as signs, as signifiers of ideas and divisions of reality. The interpretative network of signs of which the entity under investigation is a part is not a closed network. Relations are always open to reinterpretation and entities can undergo renegotiation. Object and subject of investigation cannot be easily separated. Understanding reality increasingly becomes a dialogue between the socially embedded researcher and reality. The ideas of the researcher are guided by theory, as are the means by which reality is investigated and the methodologies employed. Theories provide a framework for deciding what to study, how to study it and how to interpret the outcome of the dialogue.

GEOGRAPHICAL POSITIONS

The great continent of Asia divides in the south into three vast peninsulas; Arabia on the west, Indo-China on the east, and the southern half of India in the centre. India proper includes the whole of the central peninsula, and stretches northwards to the mountain ranges which separate it from Central Asia. The northern half is often called Continental India, to distinguish it from Peninsular India, the southern half. The Tropic of Cancer is, speaking roughly, the dividing line between these two. From this line Peninsular India stretches southwards for more than 15°, or over 1,000 miles; while Continental India extends almost as far to the north.

The natural boundaries of India are exceedingly well defined. The peninsula is separated from Arabia by the Arabian Sea, and from Indo-China by the Bay of Bengal. These two arms of the Indian Ocean give the peninsula a coast line of nearly 3,000 miles. In the north the Himalayas form an almost impassable barrier for 1,500 miles. On the north-west for 800 miles, and on the north-east for 400 miles are regions of more broken mountainous country stretching in ever diminishing altitudes from the extremities of the Himalayan wall to the sea.

These give north-west and north-east frontiers, both of which are fairly well defined. As frontiers they are imperfect only in comparison with the mighty mountainous ramparts which protect the north. No other country, of equal extent, not being an island, is so completely isolated as India, or forms so true a geographical unity. This fact, more than any other single cause, has moulded its destiny and guided the development of its people.

In strict geographical usage the name INDIA should be applied only to this well-defined geographical whole. In common use, however, a wider sense is often given to it, making it synonymous with The Indian Empire.

The Empire of India extends beyond the natural boundaries of India proper, both on the east and the west. On the east it takes in Burma and on the west Baluchistan, both of which are frequently spoken of as provinces of India. This is convenient when political matters are under consideration. But it should not be forgotten that it is only politically that cither of these provinces belongs to India. In almost all the aspects with which non-political geography is concerned, both the Indo-Chinese peninsula and Baluchistan are widely different from India proper.

On the other hand the island of Ceylon, though politically separated from India, geographically belongs to it, being a part of the great land-mass which forms the Indian peninsula. Ceylon is a " continental island," standing on the " continental shelf." At one time the shallow strait which now separates Ceylon from the mainland did not exist; and a very slight elevation of the bed of the strait would make it a part of south India again. In all their physical conditions Ceylon and the southern part of the peninsula arc one, though the name India is never used to include the Island Colony.

SURROUNDING SEAS

The Indian Seas

The Indian peninsula is washed on the cast by the Bay of Bengal, and on the west by the Arabian Sea. If the level of these seas were reduced by but a few feet, Ceylon would be

united with the mainland by a narrow isthmus. If it were reduced by 600 feet (or 100 fathoms) this isthmus would be 130 miles wide. The " hundred fathom line " is commonly taken as the boundary of the " continental shelf." The depth of the water increases slowly up to this point, and then the ocean bed drops rapidly to a depth of 1,000 fathoms or more. A glance at the map will show this. Around the peninsula of India the hundred-fathom line varies in distance from the coast from about 50 miles off Madras to 300 miles off the Gulf of Cambay. Where the coast is rocky the ocean bed often drops to a great depth with exceeding rapidity. This is the case to the east of Ceylon, where a depth of over 1,000 fathoms is reached within 25 miles of the shore.

If the level of the surrounding seas were reduced by about 100 fathoms the general contour of India would not be greatly changed. Ceylon would, as we have seen, be united with the mainland, and the whole of the peninsula would be considerably increased in width, particularly in the north. The Gulfs of Cambay and Cutch would disappear. Bombay would be 250 miles from the sea, and Karachi 80 miles; while Orissa and Chittagong would be united by land, except for one curious arm of greater depth, stretching in a north-easterly direction towards the mouth of the Ganges, which would still be claimed by the sea. But on the whole the general shape of peninsular India would be but little changed.

Far greater would be the changes wrought in the eastern and western peninsulas of southern Asia. In the west, the shallow Persian Gulf would be drained, and Arabia would no longer be a peninsula. In the east, Indo-China would stretch 1,000 miles south of Annam, and would take in the great islands of Sumatra, Java, Borneo, and the Celebes, which are all continental islands. The Andaman and Nicobar groups of islands would then each form one long and narrow island, and the two together would enclose between themselves and the mainland a deep and almost land-locked sea.

If the level of the sea were reduced by 1,000 instead of 100 fathoms, the further changes in the contour of the land would be trifling, for from 100 to 1,000 fathoms the depth increases everywhere with great rapidity. The hundred-fathom line, or the edge of the continental shelf, is thus the true continental boundary. It indicates, far more accurately than the coast line, the actual contour of the great land-mass that forms the continent.

The Natural Divisions of India

Sir William Hunter remarks that if we could view the whole of India from a balloon, we should see that it is divided into " three separate and well-defined tracts." In the north and north-west is the region of mountains, the vast Himalayan range with their allied systems. Immediately to the south is the almost equally vast region of plains, the soil of which has been deposited by the great rivers that drain the mountains. To the south again is the region of plateaux, which includes almost the whole of peninsular India. The plateaux are bounded by ranges of hills, broken along the north and east but more continuous on the west; and between the hills and the sea there is everywhere a narrow strip of alluvial land formed, like the great plains of the north, by the rivers that drain the higher land. To these three well-defined regions of India proper, we must now, if we speak of the Empire of India, add a fourth, viz., Burma—a region of alternate mountain ranges and valleys, with the great delta of the Irrawaddy towards the south.

The Himalayan Region

India, we are often told, is " bounded on the north by the Himalayas." But this great mountain chain is much more than a mere boundary. The vast system of highlands of which the Himalayas form the southern wall, is of such immense importance to India that it claims the most careful attention, and forms the natural starting point for any study of Indian Geography. From the Pamir Plateau which lies to the northwest of Kashmir, and from its great height is appropriately called in the native language, " the roof of the world," the Hindu Kush range runs in a south-westerly direction into Afghanistan. From the same centre, but running in an east-south-easterly direction, are the Muztagh or Karakorum Mountains, a range of great and sustained height. To the south of this range, and running at first almost parallel with it, is the western portion of the Himalayas proper.

The river Indus, rising in Tibet, flows in a north-westerly direction between the Karakorum and Himalaya ranges, breaking through the Ladakh Range on its course, and then, bending sharply to the south-west, divides the western extremity of the Himalayas from the spurs of the Hindu Kush. From this point the Himalayas, there called the Zaskar range, run first in a south-easterly direction, and then gradually bend round to the cast. The Brahmaputra rises near the Indus, north of the main range, and after flowing in an easterly direction for over 800 miles, rounds the eastern extremity of the Himalayas just as the Indus rounds the western. The entire range is thus held "within the gigantic arms" of these two mighty rivers. The length of the range is about 1,500 miles, and its width from 150 to 200 miles. In parts it is flanked on its southern side by low and detached parallel ranges of hills, wholly different in geological structure and history. But in most places the main range rises from the plains with considerable abruptness. Throughout its entire length one continuous range can be traced which contains most of the loftiest peaks, more than twenty of which exceed 24,000 feet in height.

The most westerly peak in the Himalayas proper is Nanga Parbat, which lies just within the angle of the Indus and rises to a height of 26,020 feet. About 150 miles to the north-east, and at the other side of the river. Mount Godwin-Austen (28,258) second, only to Mt. Everest, dominates a magnificent group of peaks in the Karakorum Range. Nanda Devi (25,661) is in Kumaon, south of the watershed that separates the Indus and the Tsan-pu. To the west are Dhaulagiri (26,826), Gosai Than (26,300), Everest, or Gaurisankar, the highest mountain in the world (29,140), Kinchinjunga (28,176),and Chamalhari (23,929). In many respects Kinchinjunga is the most notable of these great mountains. It has no rival near it, so that its mighty proportions are well seen. From Darjeeling the view is particularly fine, and is admitted by most travellers to surpass in sublimity and grandeur anything to be seen elsewhere.

The passes across the Himalayas are numerous, but of comparatively little account. Some are over 18,000 feet in height. Tlic Bara Lacha and Parang-la passes cross into Kashmir from the north-west corner of the Punjab. From Kashmir one of the best roads to Uiasa is via the Pangong Lake. Most of the trade between Northern India and Tibet has for centuries been by these routes. In the north of Kashmir the Karakorum and Muztagh passes, which cross the Karakorums respectively east and west of Mount Godwin-Austen, have been the chief routes of the trade with Central Asia. The Niti pass is north of anda Devi, near the source of the Ganges, and leads from Garhwal into Tibet. The No-la pass crosses the mountains in the north of Nepal, and the Jailep-la pass is south of Chamalhari, and east of Sikkim.

The Himalayas are, however, only the southern wall of a great mountain system which should be studied as a whole. Branching from the northern side of the Karakorum Range are the Kwen Lun Mountains which run at first due east, then bend slightly to the north, and further east to the south again. These form the northern boundary of the Plateau of Tibet. Between them and the Himalayas the elevation nowhere falls below 12,000 feet, and the average is probably over 15,000. The length of the Plateau from east to west is 1,600 miles, and its width from north to south varies from 200 to 600 miles. Its area is nearly half a million square miles. There is no other mountain-mass in the whole world at all to compare with. It is difficult to give any adequate idea of its size. The Alps of Central Europe do not cover one-thirtieth of its area, and are greatly inferior in average height. Its importance to India can hardly be over-estimated. It forms a northern rampart which no enemy has ever scaled, and with its great southern buttress, the Himalayas, exerts an influence upon the climate and rainfall of India which has done much to determine the character of the country and the development of its people.

The main chain of the Himalayas, though of much greater average height than any other continuous chain traversing the plateau, does not constitute the water- parting between India and Tibet, which is from 100 to 200 miles to the north of it.

A second wall of the Himalayas has often therefore been assumed to run north of the trough of the Indus and Tsan-pu. Dr. Sven Hedin, the Swedish traveller, claims to have traced, during his travels in Tibet in 1907-8, a continuous chain running east and west at a distance varying from 130 to 300 miles north of the Himalayas and forming the true watershed. This range he names the Trans Himalaya. It is not, however, a part of the Himalayan system, but a more ancient range, and though it forms the watershed it does not rival the Himalayas. Geological evidence shows that the general lines of drainage as they exist to-day were established long before the upheaval of the main Himalayan range, and that the process of upheaval was slow enough for the rivers to maintain their ancient channels by their own erosive power. This is seen most clearly in Nepal, where the drainage is not to the Tsan-pu in the north, but southward through deep gorges in the vast Himalayan chain, gorges which the streams themselves have cut.

Geological evidence also shows that the eastern Himalayas are far more ancient than the western, but the entire range is young in comparison with the Aravallis or the Eastern Ghats, both of which are of great geological antiquity. There was, doubtless, a time in very remote geological ages when what now forms the Indian peninsula was joined to South Africa by a broad stretch of land, of which we still have remnants in the Maldives, the Laccadives, the Seychelles, and Madagascar. In those early days the Eastern Ghats constituted the eastern boundary of the land, while the Aravallis bounded it on the west. Long ages of erosion have worn these two ranges down till they are now, probably, but a shadow of what they once were. At that time the Aravalli peaks looked out over a vast north-western sea, which appears to have been entirely cut off from the southern ocean. The rocks which now compose the mountain systems west and north of the Indus, as well as most of the great Tibetan Plateau, were then being slowly formed at the bed of the sea. There is also abundant evidence that in subsequent ages they were repeatedly thrust upwards and again submerged, existing alternately as dry land and sea bed, until the final upheaval began which slowly raised them to their present altitudes.

The Eastern Ghats were apparently also at one period connected by an unbroken chain of hills with the eastern Himalayas, a connection which continued till a comparatively recent geological age. The whole southern drainage of the Himalayas consequently, being cut off from the eastern seas, found its way across the continent and reached the sea by the Indus valley. The gradual subsidence of a vast tract, including probably the whole of Bengal as well as a long stretch to the west, established an easterly drainage and diverted the Himalayan rivers to the Bay of Bengal. But that the main drainage of the Himalayas has from the very earliest ages been towards the south is clearly shown by the character of the beds that form the Siwaliks and other parts of the sub-Himalayan range of hills. These are all composed of fresh water deposits, which were laid down by the age-long action of rivers that drained the adjacent mountains, and then thrust upwards by much later earth-movements.

Mountains of the North-West

Just as on the north the Himalayas separate India from the Plateau of Tibet, so on the north-west lower and more irregular ranges separate it from the great Plateau of Iran, which includes almost the whole of Afghanistan, Baluchistan, and Persia. The Iranian Plateau is of much inferior altitude to the Tibetan, varying from 3,000 to 5,000 feet above sea level.

It is also more broken by mountains, and in the centre contains a large oval depression, the inland basin of the Helmandi a desert tract with an average elevation of about 1,500 feet. The ranges that divide 'the plateau from India run almost parallel to the river Indus, and from 50 to 150 miles west of it. They are broken and irregular, fairly high in the north, but decreasing in altitude as they approach the sea.

From the Pamir Plateau the Hindu Kush, a flatbacked ridge of great elevation runs in a south-westerly direction, forming the watershed between the systems of the Indus and the Oxus. Branching from the Hindu Kush the short but lofty Safed Koh range runs in an easterly direction south of the Kabul river. From this range rugged and broken extensions stretch south as far as the river Gomal. South of the Gomal are the Sulaiman Mountains running north and south, of lower elevation, but culminating in the north in the lofty peak of Takht-i-

Sulaiman, over 11,000 feet high. Towards the south the folds of the Sulaimans open out, and, in steadily decreasing altitudes, bend round to the west. Further to the south again, and further west, are a number of still lower ranges, running in almost parallel ridges at first in a southerly direction, and then, like the Sulaimans, bending gradually round to the west. The most easterly of these, the Khirthar range, maintains its southern direction almost to the sea.

These various mountain ranges form the natural north-west frontier of India proper. The frontier line, as in the old Sikh days, is the extreme limit of cultivation on the eastern slopes. Theoretically this is still the boundary of British India. But partly for the sake of frontier defence, and partly to bring under effective control the turbulent mountain tribes that were a constant menace to the frontier provinces and trans-frontier trade, British influence has been steadily pushed beyond the frontier. The North Western Frontier Province lies chiefly, and Baluchistan wholly beyond it. But even the latter is now practically within the boundaries of the Empire.

Across this frontier, which extends for nearly 850 miles, numerous passes over the mountains provide gateways between India on the east and Afghanistan and Baluchistan on the west. The passes across the Himalayas into Tibet are of comparatively little moment, being only the laborious routes of a small and uncertain trade. Very different is it with the passes across the North West Frontier. Every invading host that has ever penetrated into India by land has forced its way through one or other of these northwestern passes. Through one or other of them also each of the successive swarms of immigrants who have helped to people India have found their way into the northern plains. As trade routes the north-western passes are greatly more valuable than those of the Himalayas; but their chief importance lies in the fact that most of them are possible military roads (or might easily be made such) through which an invading foe might again force his way. The safety of India on the north-west is only secured by the strength with which the passes are held. Until comparatively recent years almost all the passes that were regarded as of primary importance were situated to the south of the Kabul river. But now those which give communication between India and Central Asia via Chitral are receiving almost greater attention than any others. These are the Malakand Pass in the east of Chitral, and the Barogil and Dorah passes over the Hindu Kush. The present importance of these passes is due to the fact that they are within loo miles of the nearest Russian outpost. To the south of the Kabul River the most important of the passes are the Khyber, the Kuram, the Tochi, the Gomal, and the Bolan.

The Khyber Pass is over the eastern spurs of the Safed Koh, twenty-five miles west of Peshawar, on the road which leads from Peshawar to Kabul. This is by far the most important line of communication between India and Afghanistan. It is an important trade route and a still more important military highway. A portion of the Khyber route west of the pass is along the valley of the Kabul, but the pass itself is considerably south of the river, as in the eastern part of its course the Kabul river flows through impassable gorges. The Kuram Pass is on another route to Kabul by the valley of the Kuram river.

The pass itself crosses the western spurs of the Safed Koh at a height of nearly 12,000 feet. The Tochi Pass is on the road from Bannu to Ghazni, once the capital of Afghanistan. The road follows the valley of the Tochi, a tributary of the Kuram, and then crosses the mountains south-east of Ghazni at a height of 11,500 feet. The Gomal Pass is 30 miles north of Takht-i-Sulaiman on the road passing up the valley of the Gomal to the plateau of Afghanistan. The Gomal marks the boundary between Afghanistan and Baluchistan. The Gomal Pass is the oldest of all the passes, and has for many centuries been the route by which the caravan trade from Persia through the valley of the Helmand has reached India. The Bolan Pass, which lies to the west of the southern Sulaimans, is now traversed by a railway which connects Quetta with India, and runs beyond Quetta to the frontier of Afghanistan.

The North-Eastern Frontier and the Mountains of Burma

The eastern extremity of the great Himalayan wall is, like the western, flanked by a series of lower chains running for the most part in a southerly or south-westerly direction. To the north-east of the bend of the Bramaputra these ridges are arranged in almost concentric arcs curving round to the south, and the inner one to the southwest. Geologically they are of much later formation than the Himalayas, and their upheaval diverted the Brahmaputra from its original course to the east and turned its enriching flood into Assam and Bengal. These ranges form an effective boundary to Eastern Bengal and Assam, and constitute the natural north-eastern frontier of India proper. This frontier is, however, of little importance compared to the north-western. In the past there has been considerable Mongolian immigration into India across the mountain barriers, but India has been peopled almost exclusively from the west. And though the Burmese have occasionally raided parts of Assam, no great conquering host has ever penetrated into India on this side. Now that Burma is a Province of the Empire, the geographical frontier is no longer of imperial, but only of provincial, moment. But it nevertheless separates two parts of the Empire that are widely and essentially different.

Three main chains can be traced, which start from the bend of the Brahmaputra and continue their coarse far to the south. The chain nearest to the river bends sharply to the south-west and under the name Patkai Hills shuts in the Brahmaputra valley on the south. The Naga and Lushai Hills continue the chain, bending gradually to the south again. From the Nagas, but separated from them by a narrow valley, the Khasi Hills stretch in a westerly direction into Bengal. South of the Lushais the main range takes a south-south-easterly course, and is then known as the Arakan Yoma. It becomes narrower and lower as it passes south and bends slightly to the west again. At Cape Negrais it dips into the sea, and continues as a well-defined submerged range for over 500 miles, cropping up at the Coco, Andaman, and Nicobar Islands, and finally emerging as the island of Sumatra. The second main chain runs south from the northern highlands, marking the watershed between the Irrawaddy and the Salwin. East of Bhamo it spreads out into a broad belt of highlands, known as the Kachin Hills in the north and the Karenni Hills in the south. Further south it narrows into the Poung-loung range which divides the Sittang basin from that of the Salwin. Between these two main ranges in the south is a short and low range, known as the Pegu Yoma, It divides the basins of the Irrawaddy and the Sittang, but is nowhere over 2,000 feet in height. The third main range, the Tanen- Taung - Gyi Mountains, bounds the Salwin basin on the east. It maintains a lofty elevation further south than either of the other ranges, and, as the Tenasserim Yoma, runs down the narrow arm of the peninsula, as far as Cape Victoria. These three Yomas, or ridges, give a configuration to Lower Burma totally different from anything to be found in India. They are all fairly regular and continuous, decreasing in elevation towards the south. They divide the country into narrow valleys along which the rivers take almost parallel courses to the sea.

The Indo-Gangetic Plain

Immediately south of the great mountain wall, the general curves of which it closely follows, lies the great plain of the Indus and the Ganges. The drop from the high ranges to the plain is made with comparative abruptness, and then the plain extends southwards, with a width varying from 100 to 300 miles, till it meets the broken highlands that form the northern boundary of the Deccan. The plain is entirely alluvial, being formed of the silt brought down by the great rivers which traverse it. The thickness of the alluvial deposit does not appear to be anywhere less than 600 feet, and in the delta of the Ganges there is reason to believe that it is three times this thickness. Geologically the plain is older than the Himalayas. There are also many indications that the plain has been subjected to steady depression, acting slowly through long ages. Geologists believe that the great forces that upheaved the Himalayas also reduced the level of the plain, and that the two processes went on together, both being parts of the same great earth-movement. The eastern part of the plain was in later ages subjected to a further depression. Originally the entire plain sloped towards the west, and the whole drainage was, as we have seen, into the Arabian Sea. The later subsidence of the eastern part of the plain changed the course of the rivers, and for ages the entire drainage from the Jumna eastwards has been to the Bay of Bengal.

The area of the plain is about 300,000 square miles. It includes almost the whole of the basins of the Indus and Ganges, and thus stretches without a break from the Arabian Sea to the Bay of Bengal. The watershed between these two basins is slightly to the west of the city of Delhi, where the plain reaches its greatest height, 924 feet above sea-level. From that point to the mouth of the Indus is 850 miles, and to the mouth of the Ganges 1,050 miles. The slope towards the sea is thus extremely gentle on both sides, and the rivers consequently flow slowly, depositing their silt as they go. The northern and eastern portion of this great plain is the most fertile and populous part of India. The rainfall, especially towards the east, is ample, and the river deposits greatly enrich the soil. In the east of the plain the Brahmaputra mingles its waters with those of the Ganges, and it has been estimated that the two rivers bring down annually more than 40,000 million cubic feet of solid matter to enlarge and enrich their common delta. In the west a large part of the plain—that lying south-east of the Indus and at a little distance from the river—is comparatively barren. The soil is sandy and the rainfall scanty. From 50 to 80 miles east of the Indus the waters of the river are made available, either by irrigation works or by the annual overflow, but beyond that the land is desert.

Peninsular India

South of the Indo-Gangetic Plain a belt of highlands, fairly well defined in the west, but broken and irregular in the east, runs right across India, separating the northern plains from the plateaux of the Deccan. This belt of hills runs, roughly speaking, along the Tropic of Cancer, so that the whole of central and southern India lies within the Tropics. Starting from the west we have first of all the Vindhya Range running almost due east from the Gulf of Cambay. The Vindhyas are the lowest of the ranges that form this belt of highlands, and exceed 2,000 feet in height only in one or two places. They are separated from the Satpura Range to the south by the narrow and beautiful valley of the Narbada. North of the western extremity of the Vindhyas, but separated from them by the valley of the Mahi, are the southern spurs of the Aravallis, a low and broken range, which stretch in a north-easterly direction into the northern plain. The highest point in the

Aravallis is Mount Abu in the south-west, which attains a height of 3,900 feet. The Aravallis are, as we have seen, by far the most ancient of all the mountains in the west of India. What we see now is but the remnant of a mighty range that has survived the ravages of countless ages. The Satpuras are a shorter range than the Vindhyas, but of considerably greater height. They form the northeastern boundary of the Deccan proper. East of the Satpuras and Vindhyas, the Mahadeo Hills, the Maikal Range, and the hills of Chota Nagpur, continue the belt of highlands right across the peninsula to the plains of Bengal.

South of this belt lies the great Plateau of the Deccan, which constitutes the central core of the peninsula. Except where broken in the east by the great rivers which flow into the Bay of Bengal, this plateau maintains an elevation of from 1,500 to over 3,000 feet. It is highest in the south and west, and slopes very gradually to the north and east.

The Plateau is bounded on the west by the Western Ghats, or Sahyadri Mountains. The northern extremity of this range is separated from the Satpuras by the valley of the Tapti. From this point the Western Ghats run southwards nearly parallel to the coast, and at no great distance from it, almost to the southern point of the peninsula.

The range is fairly continuous, but there are four important breaks. Two of these are near Bombay, another near Goa, and the fourth is the Palghat Gap, 200 miles further south, where the elevation drops swiftly from over 6,000 ft. to little more than 1,000 ft. Through all these openings in the hills railways now pass, connecting the west coast with other parts of India.

In the south the Western Ghats attain to much greater altitudes than in the north. Indeed, the highest peaks to be found south of the Himalayas lie immediately north and south of the Palghat Gap, where, in sharp contrast to the Gap itself, elevations of over 8,000 ft. are reached. The Nilgiri Hills lie to the north of the Gap, the Anamalais to the south. Eastward of the Anamalais are the Palnis, and to the south stretch the Cardamom Hills, maintaining an altitude of over 4,000 ft. to within 25 miles of Cape Comorin. On their western side, throughout their entire length, the slope of the Western Ghats is fairly steep, and between the foot of the hills and the sea is a wellwatered and fertile strip of alluvial plain varying in width from three or four miles in its narrowest part to thirty miles in the south, where the mountains recede somewhat from the sea. On their eastern side the slope is less rapid, and the elevation drops gradually to that of the plateau.

The eastern boundary of the plateau is a broken range of highlands stretching southwards from the hills of Orissa till their southernmost spurs, the Shevaroy Hills, almost meet the eastern spurs of the Nilgiris. The eastern highlands, though commonly called the Eastern Ghats, have little in common with their western namesakes. They are comparatively low, seldom exceeding 3,000 ft., as well as irregular and broken, and are separated from the sea by a much broader alluvial plain. Like the Aravallis, the Eastern Ghats are the remnants of a very ancient range, worn down by long ages of "weathering." The Western Ghats, on the other hand, are of comparatively recent elevation. This, together with the fact that almost the whole drainage of the plateau is to the east, accounts for the greater breadth of the alluvial plain, which varies from 50 to 150 miles in width, and stretches far inland where the great rivers force their way to the sea.

The plateau itself is highest in the southern angle formed by the converging eastern and western ranges. In that angle lies the State of Mysore, a large part of which maintains an altitude of over 3,000 ft. In the west the plateau is almost everywhere well over 2,000 ft. From these higher levels it slopes gradually, and almost imperceptibly, in an easterly and northerly direction till it meets the Western Ghats, the low crests of which seldom rise to more than 1,000 ft. above it.

The Great Rivers and their Basins

Most rivers begin their course as mountain torrents, and this is particularly the case in north India, where almost all the chief rivers rise at exceptionally high altitudes, and before they emerge from their mountain bed have become streams of considerable volume. Most of the Himalayan rivers take their rise at elevations varying from 10,000 to 17,000 ft., and fall swiftly almost to the level of the plains through rocky channels which they have cut out for themselves. Their fall being rapid their flow is swift, and their erosive power proportionately great. Many of them flow at the bottom of steep ravines of their own making many thousands of feet deep. Mountain torrents cannot, therefore, change their course, but retain it from age to age, confined by the rocks through which they have cut their way. Such mountain torrents bring down a vast amount of solid matter to the level of the plains. The steep sides of their ravines are continually crumbling, and the disintegrated matter is washed into the bed of the torrent by which it is continually being carried down to lower and lower levels. If the rapidity of flow is checked at any point in the descent, so that a lake is formed, this solid matter is deposited, and in course of time a fertile valley is the result, through which the stream flows placidly to recommence its swifter descent further on. Many such valleys have thus been formed in the Himalayas, the beautiful Vale of Kashmir being one.

When such a river reaches the plains, it enters upon what we may call the second stage of its life. Its flow becomes slower in proportion to the flatness of the plain, and the silt which it has brought down is rapidly deposited on its banks and bed. In seasons of flood these deposits are again disturbed and carried further down. Sometimes such a river will build up its bed to a level above that of the surrounding plain. Or sometimes, having half filled up its channel, it will, in a season of great flood, overflow its banks; and as these banks are soft and easily destroyed, they are soon washed away, and the river cuts out a new, or auxiliary, channel for itself. As it approaches its mouth the flow becomes still slower, till in its estuaries it meets and mingles with the tidal waters of the sea. Here the final deposit of its solid matter takes place, and the land steadily encroaches on the sea. The old channels, or distributaries, of the river are continually being partially blocked up with silt, and in periods of flood its waters overflow and cut out new channels for themselves by means of which they reach the sea. Much of the silt brought down by such a river when in flood is washed out to sea. There it gradually settles, and the bed of the sea is raised around the river's mouth, or the numerous mouths of its distributaries. Mud islands presently appear, which in course of time are joined to the mainland, and others further out are formed. So the "delta " is always being slowly enlarged, and the land pushed further and further out.

Such is the character of almost all the rivers of India. In one point, however, the rivers of the north contrast sharply with those of the peninsula. The former are snow-fed, the latter are not. The great mountains of the north are a vast storehouse of water in the form of snow. The line of perpetual snow is at a height of about 16,000 ft. on the south side of the Himalayas, and at about 18,000 to 19,000 ft. on the north side and on the mountains of the north-west. In the winter the snow does not melt, and the northern rivers are, therefore, lowest in the early months of the year, when the effect of the monsoon rains is past, and the melting of the snow has not yet begun. But as soon as the warmer weather sets in the snows begin to melt, and then for several months there is a steady supply of water to all the rivers of the north. This is greatly increased when the rains set in, and in July the Himalayan rivers are in full flood. But the flow from the snows continues long after the monsoon rains have ceased. The rivers of the peninsula have no such supply, for there are no mountains south of latitude 27° N. high enough to reach the snow line. The southern rivers, therefore, being dependent upon the rains alone, are subject to much greater variation in volume than those which come down from the north and water the great plain.

The great rivers of the north, the Indus, the Brahmaputra, and the Ganges, drain the main slopes of the Himalayas both north and south; and the Indus and Brahmaputra bring the drainage of the north round the western and eastern extremities of the mountain chain. Some of the greater tributaries of these rivers rise north of the main chain, and make their way across the chain through deep gorges between the mountains. The Indus, its main tributary, the Sutlej, the Brahmaputra, and the Gogra, one of the chief tributaries of the Ganges, take their rise within 100 miles of each other, near Lake Manasarowar in Tibet, at an elevation of over 16,000 ft. The Ganges and another of its tributaries, the Jumna, rise in the mountains to the west of this lake. Manasarowar is thus the great hydroKraphic centre of North India.

The Indus. Rising in Tibet, the Indus flows first in a north-westerly direction for 800 miles, passing through Kashmir between the Himalayas and the Karakorums, and receiving the drainage of both these chains. Rounding the sea. Just at its bend it receives the waters of the Gilgit river from the west. Two hundred miles further on, near Attock, it is joined by the Kabul river, which, with its tributaries the Swat, the Panjkora, and the Chitral, drains the eastern slopes of the Hindu Kush and the Chitral Hills. Other important affluents from the west are the Kuram, with its tributary the Tochi, and the Gomal.

But the great affluents of the Indus are from the east. They are five in number, and, watering the Punjab, give that Province its name Panj-ab, five rivers. The Jhelum is the most westerly of these, and pours its waters into the Chenab, which, further down, also receives the waters of the Ravi. All these take their rise in the Himalayas, and, like the Indus, flow at first in a north-westerly direction. The Sutlej, the most easterly of the five rivers of the Punjab, flows from Rakas Tal, a lake to the west of Manasarowar, and, breaking through a gorge in the Himalayas north of Simla, enters the Punjab from the east. When half way on its course to the Indus it is joined by the Beas, which rises on the southern slopes of the hills not far from the source of the Chenab. At this point the Sutlej changes its course, which has hitherto been almost westerly, and after flowing for 300 miles in a south-westerly direction, joins the Chenab. One channel, called the Panjnad, thus carries the water of all five rivers to the Indus.

All these rivers being fed by the melting snows, as well as by the monsoon rains, are in flood in the late summer. After their emergence from the hills their course lie across an almost level alluvial plain composed for the most part of a soft loam. They therefore carve for themselves numerous wide and shallow channels, which they constantly tend to change. This is true also of the Indus in many parts of its course, as well as of most of the tributaries of the Ganges. In the broad plain there are no hills to confine them, and the soft earth of which their banks are composed is unable to withstand the erosive action of their flood waters.

From its confluence with the Panjnad the Indus flows midway between the frontier hills of Baluchistan and the region of the scantiest rainfall in India, the Thar or Indian Desert. It has, therefore, no further affluents of importance and in the lower part of its course gives more water than it receives. It does for Sind precisely what the Nile does for Egypt, watering and fertilizing the land for many miles on both sides. By the vast amount of silt which for ages it has brought down from the mountains it has built up for itself a bed higher than the surrounding country. Streams therefore flow from it instead of into it. This makes irrigation from the Indus for the most part exceedingly simple and easy, but at the same time it increases the danger of disastrous floods. The same force which has built the broad soft banks can also destroy them. When the river is in flood the swirling waters often make vast beaches in thoni, and widespread inundations with great loss of life and property are the result. Occasionally, also, the river will in consequence change its course, or make for itself a subsidiary channel often at a considerable distance from the main stream. At Sukkur, about 300 miles from its mouth, the river narrows into a rocky channel, and is spanned by a great railway bridge. Two hundred miles further south the Delta begins, and the river pours its waters into the Arabian Sea by many ever-shifting distributaries.

The Brahmaputra. This river, called the Tsan-pu north of the Himalayas, enfolds the eastern portion of the great mountain chain just as the Indus enfolds the western. Rising in the neighbourhood of Lake Manasarowar, it flows almost due east for a distance of almost 700 miles; then, bending sharply round the mountains, takes an almost westerly course through the narrow valley of Assam. On its way it receives the southern drainage of the Himalayas and the northern and western drainage of the Patkai and Khasi Hills. West of the Khasis it turns south, and almost immediately begins to throw off distributaries. About 100 miles from its mouth its main stream unites one of the chief distributaries of the Ganges. Their united waters reach the Bay of Bengal through two main channels known as the Pudda (or Padma) and the Arialkhan rivers. Two of the larger distributaries of the Brahmaputra, thrown off before its junction with the Ganges, flow into the Meghna, a river which drains the southern slopes of the Khasi Hills, and the western slopes of the Nagas and Lushais. The Meghna then unites with the Pudda fifty miles from the sea. Throughout its whole course from the north-east of Assam the Brahmaputra cuts for itself a multitude of channels, spreading itself in parts over a width of many miles and forming numerous river islands, most of which are covered with verdure in the dry season but completely submerged when the river is in flood.

The Ganges

The Ganges system is by far the most important river system in India. Not only is its basin the largest, but it includes the richest and most populous provinces of the Empire, provinces whose wealth and prosperity are in great part the gift of the river. The Ganges proper is formed in Garhwal by the union of two rivers, the Bhagirathi, which rises among the glaciers of Gangotri and is often called the Ganges, and the Alaknanda, which rises to the north-west of the great peak Nanda Devi and breaks through the Garhwal Himalayas. The Alaknanda is much the larger of these confluents. The river thus formed flows for 50 miles in a south-westerly direction, and then gradually bends round to the south and south-east, maintaining the latter direction till it has passed Allahabad. On its eastern and northern side it receives many tributaries, the chief of which is the Ramganga, which also rises in Garhwal. From the west and south it receives no tributary of importance till it reaches Allahabad, where it is joined by the Jumna.

The Jumna also takes its rise north of Garhwal and west of the Bhagirathi. In its course it describes a curve similar to that of the Ganges, and for the greater part of its way to Allahabad maintains a distance of from 50 to 80 miles west of that river. Unlike the Ganges, however, its main affluents are from the west and south. The most important of these is the Chambal which drains the northeastern slopes of the Aravallis and Vindhyas.

From Allahabad the Ganges flows eastwards, passing Benares, and a few miles further on is joined by the Gumti, which descends from the frontiers of Nepal. A hundred miles lower down it receives the waters of the Gogra, which rises near Lake Rakas Tal and breaks through a gorge in the mountains. On its way the Gogra receives the waters of the Sarda and the Rapti, and by the time it reaches the Ganges it rivals it in volume. Within the next 30 miles the Ganges is joined by the Son from the south and the Gandak from the north, and when north of the Rajmahal Hills it receives the Kiisi from the north.

Both the Gandak and the Kiisi rise north of the Himalayas and break through the mountain chain, while the numerous feeders of the Son drain the rocky highlands of the Central Provinces and Chota Nagpur. After passing the Rajmahal Hills the Ganges bends toward the south-east, and soon begins to throw off its distributaries to the south. The first of these is the Bhagirathi, which lower down becomes the Hooghly, and the point at which the Bhagirathi branches off is the beginning of the delta. The main stream still continues in a south-easterly direction, till south of Pabna it divides into two almost equal streams, one of which, the Madhumati, or Haringata, takes a more southerly course to the sea, and the other, the Pudda (or Padma) follows a more easterly course to Goalanda, where it unites with the Brahmaputra.

The northern and southern tributaries of the Ganges differ greatly in character. The former are fed, as we have seen, not only by the heavy rains which fall on the Himalayan slopes, but also by the melting snows. They are therefore much more constant than they would be if dependent on the rains alone. Though in flood in July and August they continue to bring down a fair quantity of water through the greater part of the year.

The southern tributaries have no snow reservoirs to draw upon. Their basins have also a much smaller rainfall than the Himalayas, particularly in the west. The ground they drain is also for the most part rocky, off which the water flows with great rapidity. The rivers, therefore, rise rapidly as soon as the summer monsoon brings its store of rain, and fall almost as rapidly when the rain ceases. When in flood they rival the northern rivers in volume, but for the greater part of the year they are little more than rivulets. For this reason they are of much less value for irrigation purposes than the Himalayan rivers. While (as we shall presently see) there are vast irrigation systems that draw their supplies from the latter, the systems dependent on the southern tributaries are few and small. The Ganges is also of immense value as the chief waterway of north India. Its volume is always sufficient to bear upon its bosom a vast host of boats of every description. At the various registering towns along its course over 130,000 river boats of various kinds are licensed to ply upon its waters. The river thus brings down a large part of the immense produce of the rich provinces which it traverses.

The Rivers of the Peninsula

Most of the great rivers of the peninsula pour their waters into the Bay of Bengal. In the map on page 28 the water-parting between east and west is shown by a line of heavier dots extending from north of the Himalayas to Cape Comorin.

It will be noticed that from the southern point this line follows the Western Ghats, till, from the northern extremity of this range, it strikes sharply westwards where the narrow basins of the Tapti and Narbada stretch far across the peninsula, the latter nearly two-thirds of the way to the Bay of Bengal. With these two exceptions the rivers which flow westwards are small and of little moment. The great rivers of the peninsula, though rising as a rule within a few miles of the west coast, make their way eastwards, gathering volume as they go.

The Narbada, rising near Mount Amarkantak, in the north of the Central Provinces, and a little to the east of Chota Nagpur, takes an almost straight course to the Gulf of Cambay. It receives few tributaries, and no large ones.

Like the Ganges it is a sacred river of the Hindus, and from its source to its mouth it is by very far the most beautiful river in India. "Of all the rivers in India," says Sir Lepel Griffin, "there is none which is surrounded by more romance and mystic interest; whilst for strange and fantastic beauty it takes high rank among the celebrated rivers of the world."

The Tapti rises south of the Mahadeo range of hills, and flow westward along the northern valley. Debouching through a gorge at their western extremity it is joined by the Piirna which drains the southern slopes of the same hills.

Like the Narbada, from which it is separated by the Satpura range, the Tapti flows westwards and empties itself into the Gulf of Cambay, a little to the north of the ancient port of Surat. The four great rivers of peninsular India which discharge into the Bay of Bengal are the Mahanadi, the Godavari, the Kistna, and the Cauvery.

The Mahanadi

The basin of the Mahanadi meets that of the Narbada, and one of its tributaries takes its rise, like that river, on the slopes of Mt. Amarkantak. The Mahanadi itself rises on the northern slopes of the hills that form the northern boundary of the State of Bastar in the Central Provinces.

It flows at first in a northerly direction till, having received its chief tributary, the Seonath, from the west, it turns to the east and flows east and south, past Sambalpur. Its numerous affluents drain a large tract of hilly country, and in the rainy season the river is of unusual volume for its length. When in flood it almost equals the Ganges. But, like all the Vindhyan rivers, it rises quickly and quickly falls again. The Mahanadi breaks through the hills by a gorge 40 miles long and of great beauty, and, after passing Cuttack, divides into the numerous channels of its delta. The river brings down a very large quantity of silt, and the delta is extensive and rapidly increasing.

The Godavari rises in the Western Ghats a little north of Bombay. It flows through the Nizam's Dominions, and for more than 350 miles its various tributaries form the northern boundary of that State. The general course of the Godavari is west-south-west for the first two-thirds of its length, then it turns to the south-west and maintains that direction till it reaches the sea. Its main tributary on the south is the Manjira, which rises in the Bhir country on the borders of Bombay. From the north, just at the point where it bends more to the south, it receives the waters of the Pranhita, a river almost as large as itself.

The Pranhita is formed by the union of three rivers, the Painganga from the west, the Wardha from the north-west, and the Wainganga from the north. Further on it receives from the north-east the Indravati, which rises on the western slopes of the Eastern Ghats and drains the unhealthy jungles of Bastar. In their passage through the Ghats the waters of the Godavari are confined for 20 miles within a deep and narrow channel, and the scenery on both sides is wild and grand. Shortly after its emergence from the Ghats it broadens into a vast and noble river, and at Rajahmundry it is crossed by a railway bridge if miles long. At Dowlaishweram, the apex of its delta, the river divides into three main distributaries and many smaller ones, and so reaches the sea.

The Kistna and its tributaries receive the eastern drainage of considerably more than one-half of the whole length of the Western Ghats. The Kistna basin, unlike that of the Godavari, is broadest towards the west, and gradually narrows as it approaches the Bay of Bengal. The Kistna rises near Mahabaleshwar ard flows first in a southerly direction. It has two great tributaries, the Bhima from the north and the Tungabhadra from the south. The Bhima rises to the north of Poona, and after flowing south-east through Bombay and the Nizam's Dominions, joins the Kistna a Uttle to the north of Raichur.

The Tungabhadra is formed by the union of two rivers, the Tunga and the Bhadra, both of which rise in the west of the Mysore State. They unite within the boundaries of that State, and, flowing in a north-easterly direction, separate the Presidency of Madras first from Bombay and then from Hyderabad. After receiving the waters of the Tungabhadra the Kistna turns to the north-east and continues its flow in that direction till it has passed the Nallamalai Hills, which cut it off from the plain of the eastern sea-board. Then it turns sharply to the southeast and so reaches the sea.

The Cauvery and its tributaries drain the whole of southern Mysore, the eastern slopes of the Nilgiris and Anamalais, and the northern and eastern slopes of the Palnis. It flows through some of the most productive and populous districts of South India, notably Tanjore, which it waters by means of extensive and ancient irrigation works. After passing Trichinopoly its delta begins. The river divides into two arms, the smaller of which, still called the Cauvery, flows almost due east, and divides again into several channels before it reaches the sea. The larger, called the Coleroon, flows in a north-easterly direction, and empties itself into the sea half way between Pondicherry and Negapatam.

A few miles from the city of Mysore the Cauvery divides, and the two arms, uniting again lower down, enclose the river island of Seringapatam, the famous stronghold at which Tipu Sultan made his last stand. Further down, two similar islands are formed, both held to be of great sanctity by Hindus. These are the island of Sivasamundram, at the southern frontier of Mysore, and the island of Srirangam, near Trichinopoly, the site of one of the largest and most famous Hindu temples.

The peninsular rivers depend for their supply of water upon the rainfall alone. For the most part also they traverse rocky land with a comparatively shallow soil into which little of the water sinks. During the dry season the river beds are, as a rule, nothing but vast expanses of sand, with a few sluggish and shallow rivulets.

But when the rains begin, these rivers rise with extraordinary rapidity, coming down from the hilly country in vast and ever increasing volume and with a swift and irresistible flow. Sometimes the "freshes " come down with such suddenness, and in such volume, as to overwhelm travellers who may be crossing their broad bed. Breaking loose from their banks these rivers often flood the country for many miles around.

But, except in their delta stage, they do not tend, like the rivers of the great northern plain, to change their course or cut for themselves new channels. When the flood subsides they sink again into their old beds. Because of the violence of their floods their action on the land is unusually great. Their flood waters are thick and muddy, and often discolour the sea for many miles from their mouths. Where the speed of their current is checked this mud, the washings of the inland hills, is deposited. Along the eastern coast this process has been going on for countless ages, for the Eastern Ghats, as we have seen, share with the Aravallis the honour of being the oldest hills in India. In the course of ages these hills have been worn down, and with their washings the rivers have built up a broad and fertile alluvial plain along the sea-board, which they are ever increasing and enriching.

The River Systems of Burma

We have seen that the mountains of the Indo-Chinese peninsula run from north to south, stretching southwards in almost parallel ranges, and in ever diminishing altitudes, from the great mountain systems of the north. Consequently, the great rivers of Burma which separate these ranges all take a southerly course. They are the Irrawaddy, with its great tributary the Chindwin, the Sittang, and the Salwin.

The Irrawaddy drains the greater part of Burma. It is a noble river, navigable by light draught steamers as far as Bhamo, 700 miles from the sea. and by smaller craft still further. Being snow fed it rises and falls, but is never very low. Till railways were constructed it was almost the only highway of commerce between Upper Burma and the coast, and it is still the chief. The Irrawaddy rises in the rugged mountains east of the Brahmaputra, and with the exception of about 60 miles after passing Bhamo, and again for a similar distance after passing Mandalay, when it turns in each case to the west, its general course is almost due south. At a distance of 800 miles from the sea it is more than half a mile wide, and it maintains a width of from, half a mile to a mile and a half all the way to its delta, except in four places where it breaks through defiles in the mountains, and, amid scenery of surpassing beauty, narrows into deep rocky channels. In one of these the river is only 600 ft. wide, but over 1,000 ft. deep. A little below Mandalay the Irrawaddy is joined by the Myitnge which drains the Shan Hills to the north-east. Thirty miles further down it receives the waters of its main tributary, the Chindwin. Rising, like the

Irrawaddy itself, east of the bend of the Brahmaputra, the Chindwin drains the eastern slopes of the Patkai and Naga Hills, and the Arakan Yoma. In its lower course, before its junction with the Irrawaddy, it waters a broad and fertile valley. A little below the point of confluence the Irrawaddy bends gradually to the south again. More than 100 miles from the sea the delta begins, and the river finds its way to the Gulf of Martaban through fourteen channels. On the most easterly of these distributaries stands the port of Rangoon, the chief port of Burma, and on the most westerly, the smaller port of Bassein.

The basin of the Sittang is separated from that of the Irrawaddy on the west by the Pegu Yoma, and from that of the Salwin on the east by the Poung-loung Hills, both of them fairly Continuous, though low, ranges, which run north and south, and enclose a rich and fertile valley from 50 to 90 miles in width. A glance at the map will show that the basin of the Irrawaddy meets that of the Salwin, about 100 miles south of Mandalay. The Sittang is, therefore, a comparatively short river, and being shallow at its mouth it is useless for navigation. It is subject also to a severe tidal bore. The tidal wave, concentrating in the apex of the gulf, rushes up the broad estuary as a wall of water, often from fifteen to twenty feet in height. The Sittang valley is fiat, and provides an excellent route for the railway to Mandalay.

The Salwin, like so many of the rivers of North India, rises amid the snows of Tibet. Bending to the south, 200 miles east of the Brahmaputra, it makes its long journey to the Gulf of Martaban between ranges of hills which in the north narrow its basin to a few miles. Throughout its whole course it has a rocky bed. At seasons of flood, when the Tibetan snows are melting, the Salwin brings down more water than the Irrawaddy. But numerous rocky rapids on its course make navigation impossible for more than 100 miles from its mouth.

Coast Line and Harbours

The total coast line of India and Burma, from Cape Monze on the western point of Sind to Victoria Point in the south of Tenasserim, is slightly over 4,800 miles in length. Seeing, however, that the peninsula of India stietches southwards from Latitude 25° N. for nearly 1,200 miles forming almost an equilateral triangle with that parallel as its base, and that Burma stretches about the same distance to the south on the other side of the Bay of Bengal, this long coast line is relatively short. It is comparatively uniform and regular, and is broken by few indentations of any magnitude. For the greater part of its length a sandy and almost level coast strip is washed by shallow seas. The waves, rolling in unbroken from the open ocean, break in the shallow water in long lines of surf, which even in fine weather are a difficulty and danger to small boats, and in stormy weather lash the shore with almost irresistible fury and make it impossible of approach. This is particularly the case on the south-east coast, but is more or less true all round the peninsula. In natural harbours India is unusually poor. Vast stretches of coast present no convenience or shelter whatever for shipping, neither land-locked bays nor navigable estuaries. India could therefore never become a great maritime country, and all her foreign sea-borne trade is carried in the ships of other nations.

Both the east and the west coasts of India are greatly affected by the surface currents, or drifts, in the surrounding seas, which are induced by the steady seasonal winds. During the south-west monsoon the currents run along the west coast of the peninsula from north to south, and along the east coast from south to north. During the north-east monsoon these directions are reversed. These currents exercise considerable erosive power, and at the same time they wash up, and move along the coast, vast quantities of sand, which is deposited wherever the force of the current is checked by its entry into a bay, or by its conflict with the river currents prolonged into the sea. The influence of these drifts is thus twofold. Where they beat upon an exposed promontory they gradually eat into the land. When they flow into a confined bay they wash more sand in than they can wash out, and slowly tend to fill the bay up. Striking examples of these two effects are to be found on the west coast, where the currents exert on the whole a more powerful influence than on the east. The extreme north-western point of Kathiawar, which is exposed to the full effect of the southwest monsoon drift as it bends round the north of the Arabian Sea, is being rapidly worn away, and the sea is steadily encroaching on the land. On the other hand, the Gulfs of Cutch and Cambay are rapidly silting up. One curious result of these drifts, seen almost equally along both coasts, is the formation of long banks of sand and mud, or bars, as they are called, a little distance from the mouths of all rivers. These bars are just beyond the scouring power of the flood water of the river, and they constitute an effectual barrier to navigation.

Beginning at the extreme west of India we have the excellent harbour of Karachi, a natural bay formed by a projecting ridge of rock and greatly improved by an extensive breakwater. Karachi is about 12 miles west of the most westerly outlet of the Indus. Being the nearest Indian port to Europe, and having direct railway communication with north India (by a bridge across the Indus at Sukkur), it attracts to itself the greater part of the sea-borne trade of Rajputana and the Punjab, and is a rapidly-growing port. Owing to the steady silting up of the entrance, however, the channel has to be incessantly dredged.

For 1 20 miles southeast of Karachi extends the delta of the river Indus, the shifting channels of which are navigable only by small craft. At one time the chief channel of the Indus discharged its waters into the Great Rann of Cutch, which separates the island of Cutch from Sind. Between Sind and the promontory of Kathiawar on the south are the Little Rann of Cutch and the Gulf of Cutch. The Ranns arc shown as arms of the sea on all maps, but they are no more than vast salt morasses, covered with shallow water only in the wet season, and in the dry months for the most part baked dry and hard.

They are the haunt of wild asses, which wander about in herds of 50 or 60, and are so timid and fleet that they can seldom be approached. The Ranns are sea-swamps in process of natural reclamation. The Gulf of Cutch is also exceedingly shallow, and when the tide is low much of it is bare sand. South-east of Kathiawar is the Gulf of Cambay, which, as we have seen, is gradually silting up. The port of Cambay at the north of the Gulf has lost almost all its sea trade, and the more famous ports of Broach, on the Narbada, and Surat, on the Tapti. are yearly being rendered more difficult of access from the same cause. Surat was at one time the wealthiest and most famous port in India.

About 150 miles south of the Gulf of Cambay is the excellent natural harbour of Bombay. It is protected by the islands of Bombay and Salsette, and offers abundant and safe anchorage. But owing to the constant deposit of silt at its entrance great care has to be taken to keep it open, and the largest vessels have to enter with caution. Bombay is admirably situated as the principal port for communication with the west, and is the chief mail route. It is almost equidistant from the north, south, and east of India, and is accessible by rail from all parts. 68. From Bombay south to Cape Comorin, though there are many small seaports, there are only two harbours that offer a safe anchorage in bad weather for vessels of large size. These are Goa and Karwar. The former is now served by the Southern Maratha Railway, and is growing in importance. The rest are "fair weather ports " only, and many of them are quite unapproachable when the south-west monsoon is blowing in force. Near the south of the peninsula there is an extensive system of backwaters or lagoon separated from the sea by broad banks of sand through which occasional openings give an outlet for flood water in the rains, and an inlet for the sea. In ancient days it is probable that these channels, as well as the lagoons themselves, were of much greater depth, and admitted ships of fair size. Now, however, though of inestim Bombay islands and Harbourable value as inland waterways, these lagoons are useless for seafaring purposes. Rounding Cape Comorin we enter the Gulf of Manar, on the western shore of which stands the small port of Tuticorin. The water is exceedingly shallow, and steamers have to anchor some miles out. but in spite of this Tuticorin has come to be of some importance as the chief peninsular port for communication with Ceylon. Northwanl are the Palk Straits, almost blocked by the islands of Rameswaram on the west and Manar on the east, and a long sand bank which almost unites the two.

On the landward side of the two islands arc shallow passages known as the Manar Passage and the Pamban Passage. Till a few years ago neither of them was more than six feet deep, and although they have recently been deepened and widened, they offer no route for large ocean steamers, which have to pass round the south of Ceylon. The west coast of Ceylon, like that of the peninsula itself, is low and sandy, and is subject to the same silting.

There are extensive backwaters, in many respects similar to those on the west coast of India, and, like them, of great value for boat traffic. The port of Colombo, nearly 100 miles from the most southerly point of the island, is rapidly becoming one of the most important ports of call in Asia. It has a fair natural harbour, which has been very greatly improved by the construction of a long breakwater. As a port of call, and the point where almost all the great steamship lines traversing the Indian Ocean converge. Colombo has superseded Galle in the south of the island, though Galle Bay forms an excellent and fairly safe harbour.

The east coast of Ceylon contrasts sharply with the west, as well as with the entire coast of the peninsula, being everywhere rocky and descending quickly to the sea. The depth of water increases with unusual rapidity, a depth of 12,000 ft., or over 21 miles, being reached within 50 miles of the shore.

There are several excellent and safe natural harbours, the chief of which is Trincomallee Bay, large enough, it is said, to accommodate all the navies of the world, and to be a safe refuge for ships of the largest order. Trincomallee is a British naval station; though in this respect it is of comparatively little moment, being off the main lines of sea-communication.

The east coast of India, from Pamban northwards to the Hooghly, has not a single natural harbour of any kind, and offers no shelter whatever to shipping in the furious storms that at particular seasons prevail in the Bay of Bengal. It is an unbroken stretch of inhospitable surfbeaten sands, or low, deltaic, mud islands. Across the mouths of all the great rivers bars stretch, which effectually close them to everything but boats of low draught. There are several ports along the coast at which there is a fair amount of shipping trade, but coasting steamers have to anchor far out to sea—in some places five or six miles and in bad weather the open sea is their only refuge. At Madras vast breakwaters have been constructed of solid blocks of concrete which enclose a portion of the roadstead. In good weather these secure for ships a smooth-water anchorage where they can load and unload without the inconvenience that they previously suffered from the restless waves. But as a place of refuge from bad weather the harbour thus made is useless, or worse than useless. The action of the surface drifts on the east coast is similar to that on the west, but as a rule not quite so powerful. In some places the land is gaining at the expense of the sea through the piling up of sand, and at others is being worn away by steady erosion. A remarkable instance of both these effects has been seen at Madras since the erection of the breakwaters. On the south side of the harbour the coast line is being pushed out to sea, while on the north side the sea is eating into the land.

Of the many distributaries that bring the waters of the Ganges to the sea, the only ones that are navigable for anything larger than river boats are the Hooghly and the Mutla. About 80 miles up the Hooghly stands Calcutta, the metropolis and premier port of India. The river is, however, exceedingly difficult to navigate, and in one part perilous.

At the bend in its course, where it receives the waters of the Kupnarayan, the James and Mary Sands have been formed. This is the most dangerous bank of quick-Long 90 of sands in the world. Vessels that strike it are swallowed up with appalling rapidity. Calcutta, therefore, holds its position as the chief port of India more because it taps the richest and most productive of all the Indian Provinces than by reason of its excellence as a seaport, in which respect it is far surpassed both by Bombay and Karachi.

In olden times the ships of the East India Company used to go no further up the river than Diamond Harbour, which still enjoys a certain amount of trade, and is now connected with Calcutta by railway. About forty years ago an attempt was made to relieve Calcutta of some of its shipping, which often greatly crowds the river, by opening a new port, called Port Canning, on the Mutla river, a channel of the delta, about 25 miles east of the Hooghly. Port Canning also is connected by rail with Calcutta, from which it is distant about 20 miles, but as a port it has not developed as was at first confidently expected.

From the Hooghly for a distance of 200 miles to the eastward stretch the low mud islands of the delta. In many parts, particularly in the east, these islands are cultivated, but they are mostly covered with low trees and shrubs, and are infested by tigers and crocodiles. The entire district is known as the Sandarbans, and is subject to disastrous floods both from the waters of the rivers and from storm waves.

One such wave which swept over the Sandarbans about thirty years ago caused an immense loss, both of life and property, no fewer than 100,000 people being drowned. None of the channels of the delta are navigable, save for river boats, till we come to the Meghna in the east, which will allow of the passage of river steamers at all times, and is an important waterway to Dacca. The Meghna is, however, subject to a severe tidal bore which makes its navigation difficult and dangerous. The Hooghly is subject to the same phenomenon, but not to so serious a degree as the Meghna.

The eastern shores of the Bay of Bengal are totally different from the western, being fringed with innumerable small rocky islands, mostly volcanic in origin. The sea bed immediately adjacent to the land being in most places rocky, the river mouths are not so greatly blocked with sand bars as is the case almost all round the Indian peninsula.

The rivers are therefore more open to sea traffic. For example, the comparatively small river Kaludan, at the mouth of which is the safe and well-protected port of Akyab, is freely open to the sea, in spite of a small bar. and can be navigated for over 50 miles by vessels of 400 or 500 tons burden. Akyab is the only port of any consequence north of Cape Negrais. Bassein and Rangoon are on channels of the Irrawaddy delta, and Moulmein is at the mouth of the Salwin. All these are easily accessible, and give safe anchorages to the largest vessels. This is also true of the smaller ports on the Tenasserim coast, Amherst Tavoy, and Mergui.

ISLANDS

With the exception of Ceylon, which does not belong to the Indian Empire and will be treated separately, the Islands of India are of comparatively little moment. They are, however, exceedingly numerous, especially off the coast of Burma, and though small, their united coast-line exceeds 3,000 miles in length. The total coast-line of the Indian Empire is 8,415 miles.

Salsette and Bombay are now connected with the mainland by a causeway, and can hardly be considered islands. Elephanta and Trombay are within the harbor of Bombay. Other smaller islands, mostly composed of volcanic rock, belong to the same group. The Laccadive and Maldive Islands are, as we have seen, remnants of the broad belt of land which, in far back geological ages, united India and South Africa. Or, more accurately, they are coral structures raised slowly by the coral polyp (which can only live comparatively near the surface of the water) upon the gradually submerging land. The Laccadives are about 200 miles west of the Malabar coast, and belong to India. The Maldives are 300 miles south-west of Cape Comorin, and are under a Sultan tributary to Ceylon. Nine of the

Laccadive Islands are inhabited, and seventeen of the Maldives, the population being respectively about 10,000 and 30,000. Rameswaram and Manar are two islands lying between Ceylon and India, the former belonging to India, the latter to Ceylon. Rameswaram is a noted place of Hindu pilgrimage. Of the many low islands at the mouths of the Ganges, Brahmaputra and Meghna, the only ones of any moment are Saugor island in the west, and Shahbazpur and Sandip islands in the east.

The islands off the Burmese coast are totally different from those to the north of the Bay of Bengal, being mostly rocky, and volcanic in their origin. They differ also from the Laccadives and Maldives, inasmuch as, with the single exception of some of the Coco Islands, there is hardly any coral formation to be found among them. North of Cape Negrais the only islands of any moment, among the many hundreds with which the coast is studded, are Barongo and Savage Island, which protect the port of Akyab, and the larger islands of Ramri and Cheduba a little further to the south. From Cape Negrais a well-defined submarine ridge runs southwards to Sumatra. About 75 miles south of the cape it crops up in the Preparis Isles, a group of minute volcanic peaks. Fifty miles further south are the Coco Islands, similar in every way except that they contain a minute volcano that is still very slightly active. Thirtyfive miles south of the Cocos the Andaman Islands begin, a beautiful and in many ways

important group, consisting of four large islands and many small ones, and stretching from north to south for a distance of over 200 miles. Further south again are the Nicobars. The Andamans and Nicobars constitute a Chief-Commissionership, and They have many excellent will claim attention later. natural harbours, well protected and with good anchorages. The same is also true of the Mergui Archipelago, which consists of many hundreds of rocky islands skirting the whole length of the Tenasserim coast.

7

CLASSIFICATION AND EXPLANATION

INTRODUCTION

This chapter outlines and justifies the different types of explanation that are regarded as acceptable within physical geography. Central to each mode of explanation are the researchers themselves. By defining the nature of the study, the questions to be asked and the entities to be studied, the researcher also sets the criteria by which a study is judged to be successful. Despite the wide variations in underlying philosophies, what is regarded as an acceptable explanatory framework needs to be established in order to understand where certain philosophies may be more appropriate for certain types of explanation and subject matter.

What is Explanation?

This question is a difficult one to answer despite its seeming simplicity. Although it is difficult to separate explanation from its philosophical framework, a general discussion of what explanation is may help to provide an outline of what physical geographers would find an acceptable explanation. Explanation can be summarized as providing an answer to the question 'why?' – or, more specifically, why is something as it is? Given what we know about why, what should we expect the something to do in the future? In answering these questions our explanations reduce uncertainty about reality and the future of our reality. Explanation and prediction are not the same, but are related to each other. Explanation could be viewed as a need to make the unexpected expected, to reduce the amount of surprise that anyone encounters. Surprise is replaced by expectation. When expectations are confounded by the unexpected, people tend to want to know why (Toulmin, 1960).

The impetus for explanation may be that simple. Some arguments could be constructed about the evolutionary advantage such a search might provide, but such speculation is not really a concern here. An argument could also be made that replacing the unexpected by the expected reduces individual and social stress. By reducing unexpectedness, explanation could also be viewed as beginning to represent reality as it really is. Stating that something has an explanation is tantamount to stating that we have been able to identify what reality is and from that identification been able to understand how reality works. Explanation and understanding of reality come together. Explanation implies that a mirror has been held up to reality and an undistorted, or at least acceptably distorted, reflection obtained. Whatever the case, the impetus immediately pushes the interaction between an individual and their surroundings to the fore. The need for explanation is based upon interaction; it is based on a need, an impulse to know about the environment and its behaviour.

The impetus for explanation need not be the same for different individuals. The example Toulmin uses is that of a stick appearing to bend in water (see Harvey, 1969). For some people this is a 'so what' moment. The significance of the interaction with reality is of no concern to them. For others, it poses questions about their reality. For the latter group of people, explanation provides an insight into how reality really is. Explanation produces a possibility of organizing reality into simple categories – what is explained and what is not. Additionally, these categories can be refined into, for example, what explanation A explains and what it does not, what explanation B explains and what it does not, and so on. In other words, a systematic body of explanations can be built up and used to define which parts of reality, or rather our interaction with reality, are explained and which are not. Viewed in this manner explanation needs to be contextualized. Without some way of classifying reality, without a particular means of interacting with reality, the sources of unexpectedness that provide the basis for explanation are unknown.

Training and experience in what to expect and what is unexpected, and the significance of each within one's frame of reference, needs to be identified. In this manner explanation, theory and philosophy are interwoven. Accepting an explanation as valid implies that there is a set of criteria by which an explanation is judged. Once again the context of explanation becomes important. What is thought to be a valid explanation tends to depend upon the rules and conventions of a particular group. This could be interpreted as saying that explanation is socially determined. The exact rules may depend on social factors, but it could be argued that the need to predict expectations of reality from explanations means that the rules must converge to provide a version of reality adequate for the purposes of that group. This could be taken to mean that rules for an acceptable explanation should be rules that permit explanation to reflect reality as it is. This would severely limit the nature of the rules that could be applied.

Hempel (1965) suggested that scientific explanation was concerned with both universal and statistical laws. Universal laws are general principles, possibly derived from observation of regularities in reality. Given an initial set of premises, universal laws are applied to these to deduce an outcome logically. This is the classic covering law model. Universal laws usually imply that we have an idea of the reasons why premises and outcomes are linked, that we have some mechanism in mind. This mechanism will always produce the outcome, given the premise. Statistical laws can be applied in the same manner to premises and outcomes; the covering law model can be used, but we may not be certain that we will always get the outcome. The laws may be statistical because we do not understand the mechanisms involved or because they reflect observations of common conjunctions of premises and outcomes. Statistical laws, unlike universal laws, are only true because we observe them with our current level of knowledge; they have the potential to change, unlike universal laws.

Causality

An important feature of any explanation is the idea of causality. At its heart the principle of causality assumes that every action, every event observed, has a cause. The event is determined by the cause. An event is as it is because it has been determined by events before it in a long chain leading back to what a researcher would identify as the cause. This means that physical geography and physical geographers tend to believe that events they observe are determined, that there is some knowable explanation, some knowable cause (or causes).

The simplest view of causality is the idea of cause and effect. An event happens prior to and probably close to another event. The first event can be interpreted as the cause of the second event – the effect. Cause and effect imply temporal and spatial proximity, although their linking may also require some ideas of why they are linked – a theory.

Simplistically, there are two views of causality: the successionist and the generative. In the successionist a cause is merely something, an event, that happens before something else. Connecting one event to another is something humans do, but the connections are merely our constructions and have no existence in reality. In this view of cause, the entities that form the events that occur in succession are passive objects. They do nothing to influence each other; they do not interact in any manner to produce events. The generative version of causality takes the opposite view. In this view an event is generated by a cause and so the two events are linked and dependent upon each other. Taken together, cause and effect could be viewed as forming a single event, they cannot be considered separately as single events. A specific cause must generate a specific effect of necessity (Harre, 1985).

In this view it is the nature of the entities themselves that generate the relationship. The nature of the entities involved in an event determine that event. An entity, by its very nature, being of a particular type, must possess particular properties that determine the manner in which it must, of necessity, behave under a given set of circumstances. Abstracting causality from the entities involved can produce information about general relationships, but it also tends to highlight the successional view of causality as cause is displaced from the actual entities involved in generating the succession.

The choice between these two views of causality has been the core of much metaphysical debate (e.g. Cooke and Campbell, 1979; Holland, 1986; Oldroyd, 1986), but the debate largely boils down to issues considered in previous chapters on the nature of reality and human ability to understand reality as it is. Hume viewed cause and effect as a succession of events linked together by humans by an assumed mechanism. This was a human construction based on experience, not a real structure of reality. Rescher (1991) views causality as part of our conceptual framework for bringing shape and order to a formless reality. The researcher is the centre of the construction of causality. A generative view of causality sees events as linked of necessity and therefore reflecting reality as it is. Entities possess the potential, depending upon circumstances, for causing effects. Causality is about real relationships between entities and their actions. The key problem is that humans have no absolute knowledge of reality, no privileged position from which to view things as they really are.

Kant tried to resolve the two views by invoking two realms of existence, the noumenal and phenomenal worlds. The noumenal world is reality as it really is. Humans do not have direct access to this world, instead all they can know is the phenomenal world, the world of phenomenon as constructed by humans. Causality then becomes a messy mix of the two. Causes and effects are generated by entities and their relations, but all we as humans can know are the succession of events. It is this succession of events upon which we impose an interpretation based on what we believe to be the 'real' relations between 'real' entities based on 'real' laws. All the 'reals' being human constructed, but constructs based upon a participation in reality, an active structuring of phenomenon. Bunge (1962) identified three meanings of causality. First, causation which associates a particular event with a particular result (cause–effect). Second, the causal principle which casts law in terms of cause and effect. Lastly, causal determinism; that is, a doctrine that asserts the universal validity of the causal principle.

The three meanings are not unconnected. Assuming that everything is explainable by cause-and-effect relationships implies acceptance of a causal deterministic view of reality. It implies that cause and effect are observable because of the underlying importance of laws in determining how reality works. Within science causality has a more restricted scope and definition. Humean cause is based on constant conjunction (Peters, 1991). Cause is always accompanied by an effect, with the cause occurring before the effect. This conjunction was believed to be sufficient to provide a basis for establishing a causal link. Such constancy in conjunction is difficult, if not impossible, to achieve in a field science, so more rigorous concepts are required. The operation of laws ensures that reality is composed of events that are causally linked both spatially and temporally: cause-andeffect relationships.

The presence of cause-and-effect relationships can be interpreted in terms of necessary and sufficient conditions. Any event could be thought of as having a set of causes associated with it. These causes are the conditions required for an event to take place. A potentially infinite set of conditions is not capable of study, so a subset or subsets of conditions are the focus of study. These are the necessary conditions and the sufficient conditions. Necessary conditions refer to the conditions that would justify the nonoccurrence of an event, whilst sufficient conditions are the conditions that would justify the occurrence of an event. Necessary conditions are more restrictive than sufficient conditions. Necessary conditions define the set of requirements that must be present if a certain event is to happen. These conditions can only be identified, however, if the event is prevented from happening by these conditions not being present. They imply an ability to manipulate reality to prevent the necessary conditions to allow their identification, a luxury that is not usually possible in field sciences.

Cause-and-effect relationships reveal the laws that govern reality. Underlying any cause-and-effect relationship is a set of structures. The operation of these structures ensures the consistency of causes and effects being observed. Where cause–effect relationships are not observed these underlying structures either are not present or their operation is influenced by other structures. Identification of the underlying structures can be useful in identifying expected causal relationships.

NECESSARY AND SUFFICIENT CONDITIONS

Simms (2002) identified a series of karst forms around the shores of Irish lakes that had a very specific distribution. The rohrenkarren or tube karren are vertical upward-tapering, closed tubes. In a detailed study of the distribution of these forms Simms found that the three lakes he used in his study, Lough Carra, Lough Corrib and Lough Mask, all shared a common feature: all were in a near-permanent state of carbonate saturation. At all three active precipitation of carbonate, as exhibited on the encrustation of plant stems and mollusc shells, was observed. Lakes that were significantly undersaturated did not show any tube karren formation. This means that carbonate saturation is a necessary condition for tube karren formation. The absence of this state means that this karst form will not be produced whatever the other favourable conditions for its formation. Identification of this necessary condition relied upon knowledge of the saturated state of other similar lakes, such as the Killarney lakes of County Kerry. Saturation was not itself, however, a sufficient condition for the formation of tube karren. There needs to be a set of sufficient conditions as well, conditions that produce the form.

Simms used his observations of the distribution of tube karren to advance a possible mechanism for their formation. Tube karren were confined to the epiphreatic zone produced by seasonal lake fluctuations. The level of fluctuation varied between lakes from 1 to 3 metres in tandem with the magnitude of tube karren found. Tube karren were formed by the condensation of water vapour in the air-filled apex of the tube. This water vapour is not saturated, unlike the lake water from which it is derived. Condensed water dissolves limestone at the apex of the tube, but dissolves less as its aggressiveness Water condensesat apexSolution decreases in aggressivenessas it runs down side of tube is reduced by reaction as it runs down the tube. This means that the base of a tube expands at a lesser rate than the apex, as limestone at the base is dissolved at a slower rate. The almost perfect circular cross-section of the tubes reflects the uniform condensation of water onto the side of the tubes and the uniform dissolution that results.

When the condensed water reaches the lake water there will be enhanced precipitation of carbonate. Carbon dioxide will be released by this precipitation that enriches the atmosphere in the tube apex. More carbon dioxide is able to dissolve into the condensing water film and so increases the aggressiveness of the water film. Associated with these sufficient conditions is a necessary condition, the absence of which would mean that condensation and subsequent dissolution could not occur. For condensation to occur there needs to be a temperature difference between the rock and the water vapour. If both the rock and the water vapour were at the same temperature then condensation would not occur. The temperature of the limestone fluctuates at a higher frequency and amplitude than the lake water. This results in the two systems being in disequilibrium, at least for a few hours, when the rock is at a lower temperature than the water.

Simms also suggests that the initiation of tube karren may require a set of necessary conditions. Calm water is essential for trapped bubbles to initiate the indentations that can develop into tube karren. For such indentations to survive the lake water must be at the point of carbonate saturation. Only under these conditions can the minor irregularities persist without dissolution to form tubes. The Killarney lakes of County Kerry, for example, had sufficient seasonal fluctuations of lake levels to permit the above mechanism to operate, but all were undersaturated so the initiation of tube karren could not occur.

Causal relationships can be represented by linkages between causes and effects. At the simplest level is the single cause-and-effect relationship. In this relationship cause precedes effect and the link is assumed to be the causal mechanism. The link itself, however, could be viewed as being composed of a series of smaller events that link the cause to the effect. Linking a rainfall event to a landslide occurring may seem like a simple cause and effect. The link between a rainfall event and a landslide could, however, be viewed as being mediated through smaller and shorter duration events, such as a rise in porewater pressure. When this series of links is observed over a number of particular instances of the phenomenon called a landslide, then the causal structure obtains some sort of stability, at least in the mind of the researcher. A problem is, however, that the causal chain perceived is capable of further and further refinement; it is capable of infinite regression, reducing causation down to the lowest level.

The causal chain may be assumed to be linear and restricted, i.e. a single causal pathway. This is unlikely in reality. It is more likely that a single event at the start of the chain will propagate a number of subsequent events. In other words a single cause need not generate a single event, and nor need the links between a single cause follow a singular path to a particular event. There may be intervening factors that produce multiple potential pathways to an event or even series of events. A rainfall event may generate porewater pressure sufficient to cause a mass of regolith to move, but it may also provide a lubricated layer over which the shear plane can move. In addition, a rainfall event may weather material in the mass of regolith and thus weaken the coherence of the mass, increasing its susceptibility to movement. The single causal event can have a multiplicity of effects, all of which can contribute to the landslide, possibly even at a range of different temporal and spatial scales.

This illustrates that once intermediate factors are viewed as part of an interrelated mesh of effects, the idea of a simple cause–effect relationship disappears. Instead, activation of different possible causal pathways becomes a concern porewater pressure and rainfall may be all that is required to explain the occurrence of a landslide, the other effects are no less real and still contribute to the timing of the landslide. A possible alternative view of cause-and-effect relationships is to view these as a causal web. Causes produce a range of effects that in turn function as a range of causes for other effects. Very soon a complex network of relationships can be built up between events with an associated range of outcomes. Tracing which of these relationships is necessary and which are only sufficient for any particular outcome becomes increasingly difficult as the complexity of the web grows.

Explanatory Frameworks

Given that explanation is the goal of investigation in physical geography then we need to look at how causal relations are put into an explanatory framework. Chapter 2 outlined the basics of deductive argument, and this style of argument is useful for some causal webs. The deductive argument provides a means of linking a cause to an effect through the intermediate stages of effects. Each stage could be seen as a logical outcome of the previous stage and so form a string of logical conclusions and premises that lead from cause to effect. Clearly, the problem of infinite regress of cause has to be addressed at some point; that is, at what level of reduction do you draw the line for an adequate explanation? Despite this problem, deductive reasoning provides a valid and formal basis for explanation. Deductive reasoning provides a basis for identifying what must be the case logically. The logical basis of the links is provided by reference to a 'law'.

The law provides the reason why the cause and effect should be linked. It provides the logical basis for the effect. If C, the cause, happens, then the law derived from an overarching theory predicts that E, the effect, must happen. Put another way, if a state of reality exists at time t0 which is knowable to the researcher, then applying law L, derived from a specific theory, predicts that at time t1 there should be another state of reality. An outcome state can be predicted from the initial state, if a law is applied to that initial state. There is a problem, however, with arguing, or rather presenting, cause and effect in this manner. Using 'state of reality' or 'state of affairs' as shorthand for the situation before and after a law has operated is very unspecific. Cause seems to be located within the phrase 'state of affairs' rather than in an object or entity. The phrase implies that had the state of affairs been different then the outcome may have been different. Questions of how different, and therefore what the role of any individual entity has within this explanatory framework, are not addressed by the phrase 'state of affairs'. Induction, according to von Engelhardt and Zimmerman (1988), refers to the situation where the controlling or initial state of affairs is known, as is the resulting state of affairs. What is unknown is the link between them, the theory and derived law that can connect the two states. Rather than arguing for a link between two states logically, induction infers the link between the states; there is no logical necessity for the inference to be true. All the problems of induction outlined in Chapter 2 are still relevant to this definition. If causation is located within entities and relations, inductive reasoning is arguing that if cause and effect are known then the link can be derived. The manner in which it is derived is unclear, and how the derived link can be assessed is likewise unclear. Any link put forward as a likely candidate can, presumably, be argued for within some context, or else it would not be put forward. If there were no recourse to information about cause and effect other than that contained in the initial statement of positions, then there would be no way in which to compare candidate links. It is only the possibility of other means of assessment, or using other information, that provides a way of assessing candidate links. Once the researcher moves into this form of analysis, however, it could be argued that they are moving from pure inductive reasoning into a fuzzier arena of investigation and explanation.

There is a third style of explanation. Often the effects are known and there is a link, a law that produced them. The unknown is the initial state of affairs, the cause. The argument works back from effect to cause, from resulting state of affairs to initial state of affairs. Unlike induction, there is a possibility of applying logical argument to the analysis. The law can be applied to the effect and a cause predicted in retrospect; in other words, retrodiction can be used as the basis of argument. Given an outcome produced by a given process, we should be able to infer what the initial situation was. This style of argument is called abduction. A key limitation of abduction is that there is no logical necessity for the initial state of affairs to have existed. Unlike deduction, the logical final statement is not a necessary one. Only if the law assumed to link cause and effect, initial and final states of affair, really operated can the argument be logically watertight. The initial state of affairs is the unknown, so there can never be certainty that the law operated – only a likelihood.

There could be other laws that could be applied that suggest that other initial states of affairs could have been the initial state. There needs to be a way to assess which of the possible links back to a range of possible initial states is the most likely. In other words, abduction requires additional information in order to help in selecting what is viewed as an appropriate explanation. From the above discussion it should be clear that only the deductive argument produces a link between cause and effect that arises out of logical necessity. There can be no other effect, given the cause and the law operating. Admittedly, unless the deductive argument is somehow linked to reality, unreal statements can be logically deduced from inappropriate or unrealistic initial conditions. This means that much like induction and abduction, deduction requires information from outside the confines of the logical structure of the argument to assess it. Deciding where, in the argument structure, to derive that information, and how that information is actually of use in assessing different explanations, is a relatively little discussed area in physical geography.

This is not to say that it is not considered and considered in some depth. The importance of sedimentary profiles for reconstructing past environments is of vital importance in Quaternary studies, but why this type of information is used and how it intervenes in the explanatory structure is not usually made explicit. Datable sedimentary markers such as molluscs are vital because they provide a temporal framework for defining when events occurred. Likewise, pollen from sediment can be used to infer the nature of past environments, providing information on the general nature of the environment at different times. Event-specific markers such as volcanic ash could be used to infer if a specific event happened and when.

Combined, this type of information can be used to test different arguments about the occurrence of specific events at specific locations at specific times. The type of evidence available may, however, focus an investigator solely on the type of events that can be identified and so limit explanation to these events alone.

In order to try to understand the importance of explanatory, or rather causal, structures in explanation in physical geography it might be useful to rethink the above discussions of different types of explanation as causal networks. Cause and law are known and the effect is a logical and necessary outcome of the two. In induction, only the cause and effect are known. A link could be made between the two, but its status is indeterminate and the number of possible links is high, if not infinite. In abduction, the effect and law are known, but the link is not stable nor is the status of cause fixed. The situation is rarely as simple as is illustrated in these three figures. Often there are a number of intervening states between the initial state and the resulting state.

For deductive explanation, this is not a major problem as long as each link can be viewed as a logically necessary outcome from the previous set of conditions. The states are linked together by the operation of the law. As the number of intermediate states increases, it may become increasingly difficult to sustain the logical necessity of the argument. Uncertainty over what the states are, over the influence of other confounding factors and the possibility of 'other' outcomes, can all affect the confidence a researcher may have in their explanatory structure. This may partly explain the concern with maintaining tight experimental control and keeping causal chains short, as this reduces the need to consider 'complicating' factors. Indeed, it could be argued that short and certain explanatory chains between initial, intermediate and final states are only possible under very specific conditions, and in relation to very specific questions. The movement for process-based geomorphology could be seen as an attempt to ensure that explanatory frameworks were based on short deductively argued chains. Although such studies might derive important general relationships between a specific set of initial conditions and a resulting state under the action of a single law, once the relationship was moved beyond the confines of its controlled system, what happened to the nature of the relationship is likely to become unclear.

Induction seems to have severe limitations as well. In the absence of other information, outside of the confines of the immediate statement of a relationship between cause and effect, the route between the two is almost completely indeterminate. There are a multitude of links that may be possible, that could be argued for, and there is no way of judging them. Similarly, a multitude of intermediate states can be suggested and a range of convoluted linkages put forward to enable even seemingly absurd explanations. As with deduction, maintaining a simple set of links may seem to make an argument more reasonable. Once there is the possibility of a range of possible links, the certainty of any single link is called into question. The effect is known and, supposedly, so is the law, the link between the effect and cause. Working back along this chain, the cause can be found. Abduction is based on being able to tell a plausible story to link effect and cause together via a valid law. There are, however, potentially a range of laws that could be applied to explain any effect. By extension this means that there are a range of potential starting points, a range of potential causes by which laws operate to produce effects. This sounds much like the problem associated with induction, the problems of how do you pin down the law and cause. Unlike induction, however, abduction starts with both law and effect; explanation, in other words, assumes that the law is the correct one to start with.

The issue then becomes forming appropriate links back to the cause. Linkages could be simple, such as a single one, or more convoluted, such as requiring a large number of linkages and the operation of co-varying laws. At this point Goodman's discussion of simplicity (1958, 1967) in such reasoning being the important guide could be invoked to aid explanation. Put simply, Goodman stated that given alternative explanations, preference should be given to the simplest. Defining simplicity is not a simple task in itself, however. Simplest does not just refer to how many links there are in a causal chain, but also to the ease of understanding, the relation of the explanation to other explanations of similar phenomena and the context within which the explanation is to operate. A long causal chain or even causal web may be appropriate in some cases where a particular law provides a simple explanation in terms of always producing a given effect, but which requires a long series of events to link cause and effect. In other words, simplicity is not an absolute property of an explanation but a contextual one. Viewing abductive explanation as being contextually bounded is important. It highlights the conditional nature of this form of explanation. Turner (2004) points out that there are two types of abductive arguments. The first is for unobservable tiny entities, the second for unobservable past entities.

Unobservable tiny entities, such as electrons, are only known through their interaction with other phenomena in experimental situations. Their existence is deduced from their effects. Their use in explanation is that they function as a unifier for a range of phenomena that would be difficult to explain without them. Additionally, they can be manipulated to produce new phenomena through which new properties about them can be deduced. Unobservable past entities cannot produce new phenomena as they cannot be manipulated. Historical entities instead serve as unifiers of phenomena only. The reasons for believing a historical entity indicates that a particular event happened, or that the entity behaved in a certain manner, and may be based on manipulation of or analogy with contemporary entities. Belief, for example, that a specific species of mollusc indicates a change from fresh- to salt-water conditions may be based on observations and manipulation of current mollusc species. This does not negate the fact that the fossil mollusc cannot be directly observed or manipulated. The presence of the fossil mollusc in a sedimentary profile can, however, serve to indicate a particular environmental condition. The fossil mollusc acts as a unifying entity integrating within it information about the nature of a past environment, even if this information is derived by analogy with current environments. Abductive argument can be biconditional or conditional. An abductive argument is biconditional if the cause thought to result in the effect can produce only that effect. The usual statement of this is that if, and only if, C then E is true. If and only if a specific cause C occurs then and only then will a specific effect E occur. There is, in other words, a necessity for a specific effect to follow only from a specific cause. In most cases, however, it is more likely that the abductive argument is only conditional. The link between effect and cause implies only a probable inference. The usual statement is along the lines: if C is true then E is true. In other words, a specific cause, C, occurs and a specific effect, E, occurs, but C need not always result in E and E need not only result from C.

There is a varying degree of doubt that C will produce E. Abduction searches for the range of hypothetical conditions that could link effect to cause via suitable laws. The laws should link the premises, the cause and effect, in a manner that ideally produces a biconditional argument. Failing this the conditional argument should be a highly probable or stable one.

At its heart abduction requires a researcher who can select. Selection of both what makes a suitable law, selection of causal pathways linking cause and effect, as well as selection of what is the 'simplest'. Abductive arguments cannot be decided without recourse to reality, without interaction to determine which of the potential pathways linking cause and effect are present and which, if any, have or are operating. In reality, as von Engelhardt and Zimmerman (1988) note, science is about the interplay of all three types of explanation: deduction, induction and abduction.

Deduction works from cause to effect, abduction from effect to cause. They work through the diagrams in opposite directions. The key difference is that deduction has a logical necessity about its outcomes, whilst abduction can only infer. Deduction permits only one certain link between cause and effect, whilst abduction admits to an infinity of links – but each link has a different degree of certainty or probability. Inserting deductive arguments into an abductive explanation could provide the basis for limiting the potential pathways linking cause and effect, particularly if induction, in the hypothesis-testing sense outlined by von Engelhardt and Zimmerman, can assess the existence of the logic outcome in reality.

Any study of causality in science is likely to involve developing an abductive framework within which competing or different potential pathways can be subject to deductive arguments and empirical testing. The exact make-up of any explanatory framework, whether it is 50 per cent deductive or more, is largely irrelevant; what is relevant, as critical rationalists have noted, is that any explanatory structure is capable of being tested at some point within its structure. The possibility of failure of a structure, the potential for falsification, remains a central feature of scientific argument, but it needs to be placed within a larger context of the logic of the intermingling explanatory frameworks of which it is a part.

Accepting the above argument, that science, or more specifically a field science such as physical geography, is about constructing explanatory frameworks that are both abductive and deductive, the next question is how can we test these frameworks? How can we decide which causal framework is correct or, rather, which is more appropriate for our needs? In the cases above there is no way of deciding which causal framework is most appropriate. Any information can be interpreted in favour of any framework. The issue can be resolved if there is an intermediate effect in the graph, in this case T. What T provides is the possibility of constructing a set of relationships in which the causal linkage of one factor can be eliminated. This is a simple illustration of what researchers have done for ages in experimental designs. Assessing the influence of one factor whilst keeping others constant is not a new idea, but within the context of causation it highlights the importance of understanding cause by intervention.

Causation is no longer seen as something that can be derived from theory alone; determining causation, deciding between causal pathways, requires intervention. Causal networks highlight that causation is a product of human intervention in a system rather than as a property existing independent of that intervention. Causation, in other words, has no meaning without human intervention. Constructing causal networks also highlights another property of causation. A set of entities and relations can be identified and defined as causal networks of a general type.

Any landslide, for example, will be expected to consist of a set of entities and relations. The occurrence of a particular landslide will involve the activation of specific pathways within the causation network. The former causal network is the general or type level, whilst the specific landslide is the token level (Pearl, 2000). The token level is an instance of the general or type level. The two levels are structurally the same, it is only the specifics of each landslide scenario that differ. The two levels interact and inform each other. More token-level events can help to clarify and refine the type-level structure, whilst the description and explanation of token-level events relies upon the structures developed at the type level.

In this manner the stability of certain causal networks is established as more token-level events confirm the structure of the type level. Once a researcher becomes confident of the nature of the type-level causal structures then prediction and extension of the explanation can take place. This may involve stripping the type-level event down to what the researcher regards as the essential components and modelling likely behaviour based on these. The constant interplay between type-and token-level structures can become ossified if the research community believes that it has clearly established the causal network. In this case measurement of an effect and its explanation may become so unproblematic that the effects are not questioned and form a building block in the development of explanations for other effects.

General and Token-type Landslides

The identification of general types of landslides relies upon an extensive knowledge of specific landslides. Detailed knowledge of the causes and characteristics of specific landslides provides the basis for abstracting information that can be used to define generalities, or rather similarities, between specifics.

A set of properties that seem to produce similar types of landslide, and upon which other types of landslide can be differentiated, can be identified. These form the basis for classification schemes such as Varnes (1958). At a simple level the classification schemes divide landslides up on the basis of the material within which failure occurs, the speed of failure and the water content of the landslide.

Using such schemes any landslide can be located within the classification matrix. Classification is not just a neutral activity. Identifying a landslide as a mudflow means that you have identified that it possesses specific properties and, in addition, that it will have other properties that you have not identified and will behave in a certain manner. A mudflow will behave as a highly viscous fluid and would be expected to move in a certain manner and at a certain velocity. The specific characteristics of any individual mudflow may vary and velocities will not be identical, but they will be expected to be within an identifiable range associated with that class. Token level or individual landslides are, however, able to influence the classification scheme. Apart from being the initial starting point for abstracting properties for the original classification scheme, individual landslides can behave in a manner that calls into question the expectations of classification. Petley et al. (2002) identified differences in velocity profiles between landslides undergoing failure along a shear plane and failure by ductile deformation.

Each landslide would have similar characteristics of moisture content, failure speed and fail in the same material. Detailed analysis of their profiles of velocity movement prior to failure highlights that the mechanisms of failure are different. In the case of failure along a shear plane, crack nucleation is followed by crack growth that establishes a failure surface. In ductile deformation the stress is such that crack nucleation is possible, but crack growth is not. This means that no shear plane can develop and the mass fails as a single unit rather than along a plane of weakness. This difference is reflected in the velocity of pre-slide movement. Identification of this difference in mechanisms of failure, based on detailed investigation of specific landslides, could be used to modify the classification scheme, the general level.

SUMMARY

Explanation is a means of reducing uncertainty about reality. Explanation implies that there is an interaction between the researcher and reality, based upon an impulse to know about the physical environment and how it behaves. Deciding what is and what is not an explanation implies that there is a set of criteria for judgement. This suggests that explanation is a communal activity based on rules and conventions. At the heart of explanation is a belief in causal relationship being present between entities in reality. Deriving where causality lies, in the succession of events or in the entities, is an important consideration. If causality resides in entities then they take an active role in determining the nature of reality.

Linking cause and effect is a difficult process and relies upon some underlying theory about how reality operates. This provides the basis for deciding that event A should be connected to event B; indeed, that event A should be the cause of effect B. From this simple relationship a different explanatory framework can be imposed. Deduction assumes that cause and the connecting law or mechanism are known and that the effect can be logically derived from these. Induction assumes that cause and effect are known, and that it is only the law or mechanism that is unknown. Abduction assumes that effect and law or mechanism are known and that a plausible cause can be inferred. Explanation usually involves a combination of all three explanatory frameworks. Deciding on the plausibility of a cause-and-effect relationship will involve the researcher intervening in reality to try to control for factors assumed to be important in the explanation.

CLASSIFICATION

The above discussion of entities and kinds provides the basis for dismissing the idea that there are natural classifications in reality. Division of reality is undertaken by researchers working within a unique interpretative context with associated versions of kinds and entities. The above discussion suggests that these kinds and entities do not correspond to reality as it is, but rather to reality as a useful framework for the researcher. Classification practices reflect this view of reality. Classification is based on usefulness to a researcher rather than determining the absolute structure of reality. Classification of reality, therefore, becomes a means to serve the researcher or group of researchers' ends. Classification is a research tool like any other; it is an aid to interpretation rather than an absolute statement about the nature of reality. Decisions about how to divide up reality, which at heart is what classification is, require the use of some property or properties associated with entities and kinds.

There are two ways in which properties can be used to classify reality: by lumping things together and by splitting reality apart. Lumping and splitting are two diametrically opposed ways of looking at reality. 'Lumping' views properties as being a cumulative means of combining entities into a single reality. Properties should be chosen that help in linking diverse entities. Commonalities are searched for. 'Splitting', on the other hand, looks for properties that enable an initially undifferentiated reality to be divided into distinct, unique parts. The two approaches to classification could be seen as bottom-up (lumping) and top-down (splitting), the former working from the level of individuals up, the latter down from the level of the morass of an undifferentiated reality. Choosing whether to lump or split can also influence the seemingly objective choices of quantification of classification. Foley (1999), for example, in relation to analysing hominid evolution, noted that multivariate statistics tend to highlight overlaps between hominid morphologies whereas cladistics operates most effectively when differences between morphologies are the focus of study.

A key question is which properties to select to differentiate reality, whether lumping or splitting. Property selection for classification may be related back to the theoretical framework used. Theory defines what properties reflect the essence of entities and which are contingent. Theory, in other words, guides the researcher and defines which properties to use for classification of the 'real' entity or kind. Ideally, properties that are essential for the entity to exist would form the basis for classification. These properties are in combination associated only with that entity and so provide a clear and unambiguous means of separating the entity from the rest of reality.

Such property identification and association is rare. Often classification requires not properties unique to an entity but properties that vary between entities. These can be measured over a range of values and entities assigned to a class based on their position in this limited continuum. Assignment requires division of the range of properties into specific zones. If a property value falls within that zone the entity is assigned to that class, if the value is outside the zone it is assigned to another.

Many properties can be used in combination to provide a more complex, composite measurement scale, or rather multi-dimensional measurement space in which to classify entities. The principles are, however, the same as for a single index. The measurement space is divided into discrete volumes. A composite value within the range of a specific volume means assignment to that class. The issue is more complex than just identifying properties of essence, however. Often properties useful for the classification cannot be measured. They may be properties that exist in theory, but which have no clear or tenable translation into measurable properties. In these cases, surrogate properties that are measurable and assumed to be related in some manner to the unmeasurable theoretical properties are used instead. The problem is that there is not usually a one-to-one relationship between measurable and theoretical properties. A measurable property may only reflect a particular aspect of a theoretical property. Similarly, a property may be illustrative of a number of theoretically derived properties. A simple measurable property may reflect a whole host of theoretical properties to varying degrees.

A lack of a one-to-one correspondence does not provide an ideal basis for classifying entities in an absolute manner. The implication is that all classification is conditional, although some classification schemes seem to be more stable than others. The Linnaean division of the living realm, as mentioned earlier, seems to be a successful and stable classification scheme. Stability does not, however, equate with reality. Just because a classification has been stable or appeared stable does not mean that we can say for certain that its divisions divide reality as it is. As with entities and kinds, classes are convenient and relevant divisions for researchers. Classification using measured properties requires another factor: standardization. Classification, particularly in the natural sciences, requires the repeated measurement and consequent consistent assignment of entities to a specific class. Classification would expect or demand that an entity be assigned to the same class by independent observers. Property measurement and class allocation need to be consistent across space and time to permit such comparison and assignment. Without confidence in the classification procedure providing consistent assignment, researchers could not be sure that they are studying and discussing the same entities and kinds. Without this confidence interpretation of phenomena would not be a focus of analysis; rather, it would be the validity of the entities used.

In this context the process of standardization of property identification, measurement and use for assignment ensures that definitional quality is maintained or guaranteed. Standardization, however, requires the implementation and co-ordination (convergence and alignment in actor network terminology (ANT)) of a complex social and material network. Indeed vast volumes of papers have been written to justify a particular classification system and the particular methods of measurement required to implement it. The example of soil classification provided below illustrates how classification schemes require detailed and often committee-based agreements concerning the precise methods to be used and the interpretation of results. Such concerns are not minor details amongst important theoretical or philosophical discussions.

Given the central importance of standardization to the investigation of physical phenomena, the life blood of physical geography, it is a surprise that it is not considered in more detail in the appropriate literature. Procedures of standardization are often so entrenched within the traditional practices of subject that they pass without comment or are relegated to methodological footnotes or dismissed as irrelevant methodological niceties. How agreed and common procedures for measurement arose, and the assumptions and often highly personalized battles that classes embody, are lost in their routine application. The discussions, the compromises, the accepted exceptions and the basis for redefinition of the standard are not part of the usual publication diet of physical geography. The sociological, and therefore, by implication, subjective and relativistic connotations are not seen as part of an objective study of physical phenomena.

Classification of Soils

Soil classification is essential to help clarify a complex phenomenon. Although most authors accept that classification is imperfect some believe that certain types of soil classification are superior to others: If, however, the criteria of grouping are based on intrinsic properties not specifically linked to the objective [of the study or activity] it is hoped to achieve they may be called 'natural'. For the broadest use of 'natural' systems of classification are likely to be more generally helpful and therefore should be based on fundamental properties of the objects in a natural state. The quote illustrates that Townsend views intrinsic properties as the key determinant of a natural versus artificial classification. This is a common view within soil classification. Inherent properties are viewed as objective criteria, yet defining what is inherent is difficult. Properties derived from the parent material within which soils form could be viewed as inherent, but these alter as the soil develops.

Assuming soils progress through predefined stages of change could provide a basis for defining inherent properties based on the development stage. This argument could become circular, however, as the stage will be defined by the properties. Another common belief illustrated by the quote is that a 'natural' classification will be superior to any other and by implication form a sound, objective basis for the definition of other, more purpose-orientated classification schemes. The implication is that somehow the 'natural' scheme proposed as the ideal has no purpose! Other authors, however, take a more pragmatic view of soil classification: [A]ny classification must have a purpose and with soils the aims can vary widely. For the most part, general systems developed for the definition of map units in soil surveys designed to aid land use will be discussed, but others have been constructed with less practical aims and more emphasis on soil formation and evolution to summarize relationships for scientific purposes.

Classification is justified as long as it is related to a specific purpose and is consistent and coherent with that purpose. Whether this same scheme will be of use for other purposes is open to debate even if the properties being used for classification are similar. This has important implications when classification schemes are used to produce products such as soil maps for agricultural productive purposes. If the classification schemes, no matter what their purposes, are referring to the same phenomena in reality, then overlaps or commonalities between schemes may be expected. These arise not only because the theory underlying identification and description of the phenomena inform both schemes, but also because theory also informs the properties to be measured and their means of measurement. This means there may be convergence in the characteristics considered in classification schemes despite their different purposes.

Clayden (1982) makes the point that there are two potential ways of classifying soils. Hierarchical classification can be used. This framework has a 'family tree' structure with an initial 'type' defining membership of the class in terms of a range of properties and finer divisions being made within these properties to refine the classification. An example of this is the USA soil taxonomy developed by the Soil Conservation Service of the US Department of Agriculture. Soils in the hierarchical system are divided on the basis of precisely defined diagnostic horizons and soil moisture and temperature regimes. Soils are then classified into orders, suborders, great group, subgroup, family and series.

An alternative means of classification is the co-ordinate classification system. In this system entities are still differentiated, but the properties used are not necessarily arranged in a hierarchy or even ranked. The former classification type is, according to Clayden, easier to remember as properties are prioritized in importance. Although the latter classification scheme has been used in Russia, East Germany and Belgium, the number of classes that are needed and their unstructured nature has meant that these systems have not been taken up internationally. Classification systems such as employed by ISRIC (1994) are the result of a long process of negotiation between usually national agencies. The first FAO/UNESCO soil map of the world began in response to a recommendation in 1960, with the first maps being published in 1969. From that date numerous meetings have been held to develop a common basis for classification, a common nomenclature and common methods of analysis.

The result is not an agreed and single coherent framework based on a unifying theory or agreed hierarchy. Instead the product is a compromise between existing systems developed for national and even regional interests and a need for international clarity in what researchers are referring to. The result is a classification scheme full of familiar local names such as chernozems and kastanozems, mixed with rigorous methods of demarcation. Clarity is provided by the use of two sets of properties for defining soils: diagnostic and phases. Diagnostic horizons are used for identifying soil units, and their presence is based upon the operation of specific diagnostic processes in the soil. Diagnostic properties are soil characteristics that do not produce distinct horizons, but which are of importance in classification. Terms such as 'andic properties' (exchange capacity dominated by amorphous material) and 'hydromorphic properties' (water saturation conditions) reflect such properties. Surface diagnostic horizons are defined so that their main characteristic properties are not altered by short-term cultivation. This is an attempt to permit the application of diagnostic horizons to lightly cultivated areas. Phases are features of the land that are significant for its use and management. These features enable land use and its impact to be considered in the classification.

Significantly, it is the soil in its 'natural' uncultivated state that is seen as the basis for classification. Soil altered by cultivation is viewed as a secondary or inferior entity with secondary properties. This reflects a general attitude in soil classification schemes, 'natural' first then modifications to the natural. Natural relates back to the various theories embodied by the different classification schemes the FAO/UNESCO classification tries to accommodate. These theories reflect a concern with understanding the causes of soil formation – primarily the role of climate. Latitudinal and altitudinal variations in climate are matched by variation in soil properties and modes of development. Deviations from these 'model' forms are viewed as oddities within specific developmental sequences, and are treated as such.

Entities, which are vital in any study, are often viewed as unproblematic and distinct divisions of reality. Traditional practices, entrenched theories and increasingly advanced monitoring technologies can all conspire to make entities appear unproblematic. Entities form the basis for the units upon which theories, and their translations to assessable hypotheses, are based. The entities studied indicate how researchers believe reality operates; how it is divided up, and by what processes divisions are made. Despite their pivotal role, entities remain relatively poorly conceptualized. Problematizing the status of entities seems to many to bring into question the whole process of scientific study. Questioning entities, however, is a key part of the dialogue between researcher and reality. What is being studied is being continually redefined, renegotiated and refined. The whole process of research results in an often slow, occasionally rapid, invisible renegotiation of an entity. The same name may be used, but the connotations associated with an entity change.

It is clear that physical reality does not change as paradigms change. It is even likely that the property-based definitions of mountains do not alter. So what did change? What changes is how the entity or individual is defined and the classes into which these individuals could be divided. Therefore a central concern has to be what do we mean by change in entities? Are we discussing a physical reality or a mental construct? Unfortunately, the answer is both. This is because physical reality can only be grasped via our theories. If these alter, what we believe physical reality to be alters – even if it appears that the entities remain the same physical things. Entities become defined and entrenched within particular theories, which in turn provide the basis for identifying new properties, that further help to define the entity, as well as guiding how the entity should be measured. In other words, new theories renegotiate entities in their own image. Physical geographers have negotiated, although they would probably generally view this engagement as improving their understanding of reality rather than as a process of renegotiation. This chapter tries to address the issues above by looking at, first, the nature of entities, what they are, and how they are defined, as well as why they are important. Of importance in this discussion is the relationship between entities and kinds. Kinds are linked to the issues of classification and standardization of reality. The act of classifying and standardizing gives an impression of a fixed, unchanging and therefore objective reality.

What are Entities?

Philosophically this is a tricky question. Entities are the units we believe exist in reality. They are the things we try to study. How they behave, and why they behave in that manner, is what we are trying to explain. Whether they really exist as we think they do, however, is a more difficult issue to clarify. We may, for example, study a landslide and try to explain its behaviour by reference to smaller parts. By focusing on the component parts, does this mean that the landslide is not really the focus of study or the basis of explanation? Are not the component parts the focus of study and so the things that really exist and hence the 'real' entities of study? Could we go even further, and believe that only physics and chemistry study the real basic entities? Physical geographers merely study aggregates of these basic entities. The next section will highlight why, within physical geography, such a reductionist view of entities, and the reality they represent, is misplaced.

A key characteristic of an individual entity is that it is assumed to exemplify a distinctiveness through having a particular property or attribute or set of these attributes (Loux, 1998). However, viewing entities as isolated islands of distinctiveness is not of great use in explanation. Entities need to be viewed as signs or as representative of something else – as illustrations of universals, for example – to be of use in explanation. There are at least two possible views of entities (Loux, 1998). One view is that reality is made up of universals. These universals are real entities that find expression, at least partially, in entities we find in our 'real' or experiential world. Such universals can be exemplified simultaneously by different objects, as there need not be a one-to-one correspondence between universals and experiential entities. Universals define the properties a thing has, and its relationship to the kind to which it belongs. Mudslides have a set of properties by virtue of being mudslides.

These properties are universal, possessed by any entity defined as a mudslide. Mudslides we observe are not 'pure', however. The realization of the full set of properties may be polluted by the nature of the material the mudslide occurs in, the specific environment conditions it occurs under, and so on. We know that these properties are not fully expressed because we have in mind the ideal type, the universal mudslide, which exists in reality but only finds expression in the pale realizations we can observe.

Another view is that reality is only made up of particulars, of individual entities themselves. These entities are not examples of some underlying universals. Instead, each particular has properties that may or may not be the same as properties possessed by other particulars. In this view two things may occur that possess the same velocity of movement, that are composed of roughly the same material, and seem to behave in a similar manner as far as we can identify and measure their behaviour. They are both called mudslides. These mudslides have similar properties because we say they do.

Possession of similar properties is not an indication of any relationship between the two entities, of any prediction of behaviour of one from the other. This possibility would imply that both entities are linked in some manner to a universal. Both views of reality do, however, accept that entities can be structured and complex in nature, and not the simple basic units of reality a more reductionist view might envisage. Properties that are the hallmark of a distinct entity can, however, be held in different ways. Any entity could be viewed as an entity shell devoid of any properties – the so-called substrata view. Properties are independent of the entity and not necessary for its existence.

An entity exists and properties fill it. Despite being associated with an entity, the properties do not define it. Almost diametrically opposed to this view is the concept of an entity as a bundle of properties. There is no empty shell to fill. Instead, it is the association of properties or attributes that define the entity. An entity has no physical existence outside of the potentially contingent association of properties. Without properties the entity melts away; it has no substance. Both these views have severe metaphysical problems, and within physical geography neither would be able to enhance explanation. The first assumes the existence of an entity without properties, the very things we identify, define and by which we measure an entity. The second provides no solidity to a form, just viewing an entity as, at the extreme, a contingent and potentially random association of properties. An understanding of reality based on either would be very limited.

Entities and Kinds

A way out of the explanatory dead-end of the above views of entities is provided by the concept of substance or essence. The idea has been traced back, as much of philosophy, to Aristotle. The central tenet of the idea is that entities can be seen as irreducible. Irreducible entities are not merely the sums of their parts, they are more than this. When put together the whole has properties grounded in universals that refer to the essential nature of the substance of the entity. It is these properties that define an entity. This means that not all properties associated with an entity are of equal importance for the essence, the substance of the entity. Without a limited subset of specific properties the entity could not exist or be identified. These properties define, or rather permit the researcher to identify, the substance of the entity. The properties are not, however, filling an empty shell.

The properties exist, take the values they take, and enter into the relations they do, because they are part of a specific entity. The relationship between entity and properties is mutual and indissolvable. A key universal is that of a kind. Kinds provide an identikit of the essence of an entity. Membership of a kind defines those properties that should be present in any entity of that kind. Non-essential properties do not alter an entity's membership of a kind, but they can mean that entities of the same kind need not all have properties in common, nor essential properties with the same values. Kinds can be thought of as templates, outlining the basic contours of what it means to be an example, a member of a kind. Kinds can also form nested hierarchies of more general or more specific essences, depending on which way you move through the hierarchy. Animals, for example, could be seen as a general kind of which apes are a more specific kind and humans an even more highly specific kind. Each more specific kind retains the general properties of the more general kind, but is differentiated by other properties which do not contradict or prohibit their membership of the more general kind. Linnaean classification in biology could be seen as a classic example of nested kinds.

Providing a template for the essence of an entity is fine, but does it aid explanation? If you had an entity of a certain kind you would expect it to behave in a certain way. Membership of a kind defines the relationships possible, the essential properties, and, potentially, how an entity can change. But is the kind then the appropriate focus of explanation? Could not a geomorphological feature, such as a slope, be composed of other smaller kinds such as soil? Are these smaller kinds more fundamental than the slope? Are they more appropriate as foci of explanation? Could you keep this reductionist regression going until you reach 'real' kinds, often referred to as natural kinds? Rhoads and Thorn (1996b) highlight the significance of the object rather than the method of study for geomorphology. They discuss whether landforms are natural or nominal kinds. Natural kinds they view as having some objective and real nature in reality, whilst nominal kinds are humanly constructed artefacts.

If natural kinds exist, their objective nature could imply a superior status in any explanation. Explanation involving natural kinds would be viewed as involving real things, reality as it is, rather than a potentially subjective and changeable human artefact. Studying an entity as a natural kind has distinct advantages. Researchers can assume that they are studying the world as it really is and so obtain objective, independent information. Investigation of natural kinds derives the essences of 'real' objective entities, their generative mechanism and causal powers. Powers of an entity exist independent of its context and who is studying it.

The identification of a natural kind from other kinds becomes important for explanation if only natural kinds are seen as really existing. Establishing some criteria or practice for demarcating natural and human kinds becomes a vital task. Putnam (1973) and Kripe (1980) claim that empirical research provides the basis for naming and assigning natural kinds. Li (1993) suggests that this merely moves the question of kinds from the realms of philosophical debate to scientific debate without really tackling the basis of the question. Schwartz (1980) similarly holds that natural kinds are defined by traits that are discoverable empirically and tested for by his counter-example test. Scientists identify which traits are associated with natural kinds through study of individual entities. These entities are, however, studied in a manner consistent with their being the member of that kind. This somewhat circular argument makes it difficult to break out of an empirical bind.

Unfortunately, for physical geography the status of entities such as rocks, rivers, and even the events that trigger landslides, as natural kinds, is debatable. Aristotle, for example, only gave the status of natural kinds to a limited range of entities (Loux, 1998), the elements of his day. Contemporary fundamental quantum physics might claim the same status for its entities, but the changing nature of what has been seen as fundamental or elementary entities over the last 2,000 years seems to count against a view of a once-and-for-all answer to the delineation of a fundamental 'natural' kind. Although the existence of geographic entities is not denied, except in extreme interpretations (e.g. van Inwagen, 1990), their explanatory status is often viewed as inferior.

Dupre (1993) and Shain (1993) suggest that the definition of natural kinds is dependent upon the context of enquiry. Li (1993) argues that in naming a natural kind there is always a vagueness at what, precisely, the process of naming is directed. It is only possible to know natural kinds through instances, individual entities that exemplify the kinds. There is no direct access to a natural kind. Naming is based upon experience, both individual and collective, and not upon some absolute knowledge of reality. Identification and naming is based upon a comparison of entities. As the process of comparison can be virtually endless, it is for all practical purposes never complete or capable of completion. This means that the absolute assignment of an entity to a kind can never be certain: I have argued that, because we have no direct access to a natural kind as we do to an individual object, when we name a natural kind exactly what the kind is cannot be determined without further focus. The process of further focusing can never come to an end.

Nominalists, according to Hacking (1983), do not deny the reality of an external world containing entities which interact. They believe, however, that we impose our classification, our own divisions, upon this reality. Hacking (1991) suggests that kinds become defined and important when their properties are important to the individuals who want to know what entities and, by extension, kinds do and what can be done to them. Practice and experience become central and important factors in the process of identifying kinds: When we recognize things to use, modify or guard against, we say they are of certain kinds. Singular properties are not enough. Realizing that a thing has some properties or stands in certain relations prompts belief that it is of a certain kind, i.e. has other properties or stands in other relations. Kinds are important to agents and artisans who want to use things and do things. Were not our world amenable to classification into kinds we cognize, we should not have been able to develop any crafts. The animals, perhaps, inhabit a world of properties. We dwell in a universe of kinds... Natural kinds, in short, seem important for homo faber.

Hacking (1999) identifies two types of kinds that may prove a useful typology in geography. Indifferent kinds have no reaction to being named. A limestone rock does not respond to being classified as a limestone rock. Its nature or essence does not alter, nor does it begin suddenly to behave as a limestone rock when it did not before. This is not to say that we, as investigators, do not interact with the rock in a particular manner because we have defined it as a limestone rock. A stonemason might work the limestone in a particular way and expect a particular response. Scientists may test for the presence of limestone by watching for bubbling from the rock surface after the application of hydrochloric acid. Neither of these behaviours provokes a response from the rock independent of our interpretation and action upon it. Interactive kinds, on the other hand, react to the act of classification. An individual defined as a refugee becomes embroiled in a set of relationships that operate because of that act of classification. The individual reacts to that classification and responds in a manner permitted by or in keeping with that classification.

The act of classification provokes a response. Most, if not all, kinds in physical geography could be thought of as indifferent kinds. There may be an implicit assumption that indifferent kinds are somewhat more 'real', more stable than interactive kinds and therefore a 'better' basis for explanation. Even indifferent kinds, however, are both social and real (Hacking, 1999). The basis for identifying and researching kinds is the dialogue between reality and the researcher. Extraction of properties is always made for some purpose. The distinction is not between some real, indifferent natural kind and some interactive socially constructed kind. There is no distinction to worry about. Trying to establish a distinction is irrelevant. Goodman (1978) identifies relevant kinds as those defined as relevant for some purpose by the user. Goodman emphasizes that relevant kinds assume an external reality with real entities. This reality is, however, only accessible through our actions, guided by our intents. We divide reality up and make entities for our own relevant ends. There is no indication, and no way of knowing, if our divisions of relevance match 'real' divisions.

Kinds in physical geography are always relevant kinds derived from a time- and place-specific dialogue. Their contextual nature makes them no less real in the framework of study of which they are a part. Semiotics, as put forward by Baker (1999), has a similarly blurry image of entities and recognizes their indeterminate status. Entities have to be viewed within the web of signs of which they are a part, as noted in Chapter 2. Entities, and by extension kinds, become contextually defined rather than objects defined independently of study context: As a thing it merely exists, a node of sustenance for a network of physical relations and actions. As an object it also exists for someone as an element of experience, differentiating a perceptual field in definite ways related to its being as a thing among other elements of the environment. But as a sign it stands not only for itself within experience and the environment but also for something else as well, something beside itself.

It not only exists (thing), it not only stands to someone (object), it also stands to someone for something else (sign). And this 'something else' may or may not be real in the physical sense;... Divisions of things as things and divisions of objects as objects are not the same and vary independently, the former being determined directly by physical action alone, the latter being mediated indirectly by semeiosy [sic], the action of signs . . . Divisions of objects as objects and divisions of things as things may happen to coincide... But even when they coincide the two orders remain irreducible in what is proper to them. This quotation provides a useful perspective on the semiotic or pragmatic realist interpretation of entities. An entity takes on a meaning and is only constituted because of its relations in a triad of sign, interpretant and interpreter. The entity exists as a thing,

Species as Natural Kinds

The status of species as a natural entity (Ruse, 1987) for biological classification has been questioned by a number of researchers. Dupre (2001) suggests that the species do not effectively function as a unit for both evolution and classification in general. He makes the point that a unit of evolution needs to be an individual, whilst a unit of classification by definition has to be of a kind. It is the individual organism that is in competition with others of its kind, as well as eking out a living from the environment. According to Dawkins (1978), the unit of evolution is the gene that provides the basic building block for evolution to occur. Organisms are merely differentiated shells surrounding genes. The interdependence of organism and gene is just one problem with this restrictive view of the basic unit of evolution. According to Dupre (2001), evolution occurs when a set of properties that characterize individuals in a phylogenetic lineage change over time.

The lineage is the unit of evolution, not some entity at a single point in time. If this is the case then what properties alter and how does their alteration not change the essence of that species or gene? Both for species and genes the temporal dimension raises problems for identification and classification. The idea that a species can be the same thing throughout its evolution implies that the same species, or rather aspects of it, exists at all stages of its evolution. Such a view implies that each current species is somehow inherent in its past evolutionary lineage. Levins and Lewontin (1985) and Lewontin (1995) highlight that the organism may be the unit of evolution, but it is not necessarily a passive unit. A classic view of evolution implies that the environment sets problems and organisms find solutions. The environment is fixed; organisms are malleable. The organism is a passive respondent to changes in the environment. The advent of genetics may seem to imply a more active role for the organism, but instead it has emphasized the internal, inherent characteristics of the genes. The organism is still a passive unit; the genes alter so the organism has to. The organism becomes almost an infinitely deformable conduit for the external influences of the physical environment and the internal influences of its genetic make-up. Levins and Lewontin (1985) and Lewontin (1995) instead suggest that the organism is an active motor of change and, at least partially, a determinant of its own destiny. They base their claims around a number of key points.

First, organisms determine what are relevant characteristics of the environment for them. The concept of a simple niche model to reality is difficult to sustain. Adaptation to fit a niche requires defining the niche before the organism exists. In a similar vein, organisms can create their own microenvironments. As Levins and Lewontin (1985) note, even without manipulation of their surroundings, all terrestrial organisms have a boundary layer of warm air resulting from the metabolism of the organism. Second, organisms alter the external world that they interact with. Plants alter the substrata they are on by their activities, and alter the soil structure and chemistry locally. Third, organisms interpret the physical signals they receive from the physical environment. Any environmental change is mediated to the organism through its senses, and its response will depend on what it senses and how it interprets this change. Fourth, organisms can alter the statistical pattern of variation in the physical environment. Changes in air temperature may be dramatic, but an organism with feathers does not sense anything other than extreme changes in temperature.

Lastly, it is not the environment that defines the traits to be selected by evolutionary processes, it is the relationship between the organism and the environment that defines these. The organism is an active participant in the environment, and it is how the organism relates to the physical environment, other organisms and its own species that determines which traits are of use and which are not. There is no predefined set of traits that are inherently better than some other set of traits; it all depends on context. The above points illustrate that the unit of evolution is difficult to isolate from its relations. Likewise, the organism as a member of a kind needs to be thought of as an active rather than passive agent. Its properties are context-dependent and subject to change, but these properties are present, and therefore have the potential to change, because the organism is a member of a particular kind. Activation of these properties is not inevitable, it all depends on the relations an organism has in a specific context.

The above discussion implies that classification of species is fraught with problems. Dupre (2001) suggests that classification, when associated with a central theory, can produce a 'real' classification of reality, even if it is, he believes, a weaker claim that is traditionally assumed for natural kinds. Kinds themselves, however, are not the sort of things that exist in reality. Instead it is the members of the kind that undergo the effort of existing in reality.

MAGNITUDE AND FREQUENCY - ENTITIES OUT OF CONTEXT

Wolman and Miller (1960) outlined the significance of magnitude and frequency of events in geomorphology by abstracting events from their context of operation. Measuring an input, an event, to the system and then its immediate output or effect, they could construct a relationship between the two that was meant to hold for any system anywhere.

Events were defined as discharges of given magnitudes, whilst output was defined as the sediment load carried by the river. The theory they developed, or 'truism' as Haines-Young and Petch (1986) describe it, was that middle-size events, normal events, did the most work in a catchment. Small events were frequent, but of insufficient power to move material. Large events moved a great deal of material, but were not frequent enough to move material with sufficient regularity to sculpt the landscape. It was the middle-size events (bankfull discharge events in their study) that occurred with sufficient frequency and force to move material around and out of the catchment so as to sculpt the landscape.

Lane et al. (1999) noted that this initial description of the relationship between magnitude and frequency relied heavily upon the idea of a 'normal' operating mode for the fluvial system within a catchment. The idea that the fluvial system tended towards an equilibrium form with a 'normal' mode of operation was an important basis for legitimating the abstraction of events and effects from their context. Abstraction was permitted, because the events and effects were illustrations of this normal mode of operation. Event characteristics were reflections of an ideal event type that had a specific and contextless relationship to an equally ideal effect. This meant that the relationship between event and effect could be modelled as a simple input–output relationship that could be transferred between fluvial systems. The magnitude–frequency curve derived could be applied to any system regardless of its environmental context. Subsequent development of the magnitude–frequency concept by Wolman and Gerson (1978) gave some recognition to the context dependence of the abstracted relationship. They introduced the idea of effectiveness to describe how different landscapes may respond differently to the same event, the same input.

Effectiveness related the magnitude of the change caused by an event to the ability of the landscape to restore itself. Both parts of this equation were context-dependent. Within a temperate environment an event of a given magnitude will have an impact of a specific magnitude. The temperate environment will tend to recover from this impact and move back towards its 'normal' state. In a semi-arid environment, the impact from an event of the same magnitude will be greater and the time required for recovery will be longer, possibly even longer than the time intervals between events producing the same impact. Although context is now introduced into the analysis, it is still a very limited view of context. Likewise, the system still retains a movement towards equilibrium or 'normality' as its functional goal.

Lane and Richards (1997) highlight the need to distinguish between the 'immanent' ahistorical processes and the 'configurational' (Simpson, 1963). The former are general processes that always occur under specific conditions, whilst the configurational represent the result of the historical interaction of the immanent with particular historical circumstances. This mirrors the complicated relationship between event magnitude–frequency and impacts. The relationship can be abstracted and general properties derived. The relationship is a generalization, often a statistical caricature of complicated and temporally distinct relationships. The relationship cannot be understood as an ahistorical process. Generalities are only derived from particular circumstances. Without knowledge of these circumstances the generalities cannot be related back to reality. General properties are only discernible because the processes that cause them operate within specific contexts derived by unique historical sequences of change. an object and sign simultaneously, and each of these elements is interdependent and uninterpretable without the others.

The sign, the focus of study in semiotics, is the 'something else' of the quote. The sign is constructed through interpretation by an interpreter of an object. If the object and thing are identical the sign is an interpretation of reality as it is. The coincidence can never be known for certain, but it would also be inappropriate to assume its presence meant an object could be reduced to a thing or even vice versa. The nature of a thing is associated with its relations to a physical reality, relations beyond the knowable for a researcher. The thing is wholly or partially related to an object that is constituted by relations to the researcher. This latter set of relations means that the object is constituted via signs; it stands as an outcome of interpretation. This means that its very nature is different from that of a thing, the reality, that it is a representation of. Any knowledge a researcher can glean about reality is via the object and so only knowable via signs, via interpretation. Knowledge without relations to signs is not possible.

Entities are not reducible to some common baseline of reality, they are only reducible to objects of study, with all the difficulties of being part of a dialogue that this implies. In some ways this interpretation of the individual entities is similar to the relational view of entities put forward by Whitehead (1929). He understood entities to be capable of distinct definition, but only by virtue of an entity's relations to other entities and, in turn, their relationships to the wider whole of reality. Any single entity at a specific time contains within it, its relation to itself in the past and future. Past relations express the constraints on an entity, whilst future relations, as yet unrealized, express the potentialities of an entity. In Whitehead's view all the relations that come to define an entity in the present are the constraints of its past and explain why it is as it is. The uncharted future of an entity is not random nor completely expansive. An entity's future is constrained, but not singularly determined by its past (what it is).

The nature of an entity defines its possible futures, its potential relations. In this way, an entity embodies both what it is, what it has become and what it could be or is becoming. There is, however, no need for an entity's nature to be rigidly fixed or stable over time. Entities can vary in their spatial and temporal stability. Whitehead's vision of entities as observer-defined abstractions, solidifications, of the flow of processes, captures the distinct, but ever-changing, nature of an entity. Entities require flows of processes, their relations with other entities and the wider whole of reality to continually reproduce themselves. Stability of these relations provides the appearance of solidity, of illusory permanence. Semiotics, with the continual creation and destruction of networks of signs that constitute an entity, is a view in a similar vein.

The semiotic approach means that both entities and kinds do not need to have any physical reality, although they will have relations to this reality. In semiotics and pragmatic realism there are no natural kinds or entities – neither exists independently of an interpreter: And the metaphysical categories that emerge through this endeavour yield an understanding of nature or the dynamical object which, as an indefinitely rich evolving continuum that must be made determinate for our awareness by the manner in which we 'cut' into it, cannot provide the basis for final ultimate knowledge. Nature or the dynamical object, with its qualitative richness, lawful modes of behaviour, and emerging activities, constrains our interpretations, pulling them or coaxing them in some directions rather than others. It answers our questions and determines the workability of our meaning structures, but what answers it gives are partially dependent on what questions we ask, and what meaning structures work are partially dependent upon the structures we bring.

Thus, within this interactive context of interpretation and constraint, different structurings yield different isolatable dynamical objects, different things, different facts. Truth is always related to a context of interpretation. This is not because truth is relative but because without a context of interpretation the concept of truth is senseless, indeed literally so. Knowledge involves convergence, but convergence within a common world that we have partially made, and continually remake in various ways.

As the quotation points out, the divisions of reality, the kinds studied, are of our making. We do not, however, start with a blank page. All our interpretations are constrained by an underlying and independent reality. We interrogate that reality and kinds emerge as an outcome of that interrogation, but are not existent independent of that interrogation. Entities of study, likewise, are dependent upon this interrogative process. It is of vital importance, however, to realize that the dependence of kinds, entities and objects upon interpretation does not mean that they reflect a purely relativistic version of reality. Kinds and entities are not relativistic in the sense that there are an infinity of other possible interpretations, all equally valid, as there is no basis for assessing the validity of each. The basis for assessing the validity of entities and kinds constituted is the interpretative context of their constitution. It is only relative to this context that appropriate methods of validation can be constructed. It is the interpretative context within which judgements concerning appropriateness can be passed.

The interpretative context, like the entities and kinds within it, is open to continual renegotiation, to continual alteration and evolution. Many scientists would claim that this renegotiation is the convergence mentioned in the quotation towards a 'truer', in the absolute sense, version or interpretation of reality. This assumes a progressive direction to the alteration of the interpretative context of which we can never be certain.

The continuation of certain entities and kinds with changing interpretative contexts testifies to their adaptability rather than proof of their ultimate reality. A more appropriate metaphor for alteration of interpretative contexts may be that of evolution. Changes in interpretation in the interrogation are appropriate for the needs of the researcher in that time and place. These needs, and the alterations they cause, constrain the nature of further alterations in interpretation; but how is unclear, because, in the present, we work only with the singular, historically unique and constrained outcome of the past.

SUMMARY

Entities are the centre of any study of reality. Entities are the units that we believe exist and which we can manipulate and use for testing ideas. Entities have properties or attributes associated with them, but not all of these are essential for defining the essence of an entity. Entities and kinds are intimately related. Kinds provide a template for what it is to be a member of that kind. They dictate the essential properties that define a member of that kind, as well as the behaviour to be expected. The kind becomes the focus for explanation; the entity becomes an illustration of the kind. There are different types of kinds and it has been argued that research should focus on natural kinds. Even if geographic natural kinds exist, all kinds are defined for a purpose. This means that all kinds are to a large extent observer-defined and defined for a purpose, not as some natural building block that forms the 'real' basis for explanation. Kinds and entities are enmeshed within a network of signs and their meanings are likely to evolve as their purpose alters. Hacking (1999) identifies two types of kinds: indifferent and interactive. Physical geography deals with indifferent kinds; that is, those that do not respond to the act of classification.

8

SYSTEMS ANALYSIS IN PHYSICAL GEOGRAPHY AND SOCIETIES

Systems analysis came to the fore in physical geography in the 1960s and found formal expression in the classic textbook of Chorley and Kennedy in 1971. The introduction of systems analysis into physical geography was not without critics, but the new approach rapidly became one of the cornerstones of thinking about the physical environment. The continued success of systems analysis is probably best illustrated by a perusal of any set of modern textbooks on introductory physical geography. Often the titles of such texts use 'systems' as an explicit indication of their approach. Almost invariably the contents pages divide the subject matter into specific environmental systems, each of which is considered in turn – for example, the atmospheric system, the lithospheric system, the biosphere, and occasionally with a chapter attempting to integrate the disparate systems at the end of the text. Systems analysis, if judged by column inches of text in undergraduate books and even research papers, has become the overarching framework for understanding in physical geography. With such success it is important to understand what systems analysis is. Systems analysis has also moulded thinking about the physical environment. Entities and relations are viewed in a specific framework and studied according to expected modes of system behaviour. An understanding of the constraints on thinking imposed by systems analysis needs to be clear, so that the limits of systems analysis do not become a barrier to comprehension of the physical environment. Systems analysis did not develop in physical geography – its pedigree is far longer. 'Systems', as a term, has been around since modern science developed in the seventeenth century. The rise of systems analysis owes a great deal to the attempt to develop an integrated and all-encompassing framework for all science in the twentieth century. The existence of such a framework implies, first, that all reality is capable of being understood; there are no areas or topics outside of its analytical scope. Second, all reality can be understood in a common framework using the same sets of terms. This means that understanding in supposedly different subject areas does not require specialist terms or specialist knowledge, but rather translation of these terms to the common terminology of systems analysis. Third, as there is a common framework, all reality can be expected to behave as predicted by this framework. All reality becomes potentially predictable and, by implication, potentially controllable. Systems analysis should not, however, be viewed as something that emerged fully formed for incorporation into physical geography. Similarly, systems analysis has not remained a static form of study in the 40 or so years it has been used within physical geography.

Having said this, the basic tenets of systems analysis remain pretty much the same as they did 40 years ago and the implications of the approach for how the physical environment is studied have become ingrained within geographical practice. Van Bertalanffy proposed general systems theory as a unifying framework in the late 1930s. Tansley pushed the concept of the ecosystem in the same period. Both had a similar view of a system as an integrating concept unifying the different entities found in an environment within a common analytical framework.

The advent of cybernetics in the 1940s and 1950s added another level of understanding to the systems approach (Shannon and Weaver, 1949). Cybernetics is an underrated influence on the development of systems analysis within physical geography. Physical geography tends to focus on its perceived antecedents within ecology, biology and physics as the source of its development of systems analysis. As such these subject areas hark back to the pre-war era of geographic research. The work of Gilbert (1877, 1896) or Penck (1924), for example, is reinterpreted in system terminology. Gilbert's use of terms such as 'system', and his systematic approach to analysis, is taken as an indication of systems thinking even if the terminology had not been invented. Likewise, Penck's view of landscape development is often recast as a systems diagram (Thorn, 1988) even though the term was never used by him. Both examples reflect, as noted in Chapter 1, a tendency to reinterpret the past in the light of current thinking. The influence of these individuals and subject areas is not denied, but the nature of the systems they suggested was not the same as 'systems' as used and understood today.

Cybernetics, the study of self-regulating mechanisms in technology, along with the development of information theory, had a profound influence on the nature of systems analysis that physical geography encountered in the 1960s. Shannon and Weaver's development of a mathematical basis to information flow and interpretation provided a means of quantifying change in abstract phenomena. Coupled with the lexicon for describing relationships between entities provided by cybernetics, these new analytical frameworks resulted in the development of a highly mathematical and formalized description of systems analysis by the 1960s. It was this version of systems analysis that physical geography tried to bring into the heart of the subject. The success of systems thinking cannot be denied within physical geography; the total incorporation of a fully fledged systems analysis is harder to justify.

The key ingredients of a system are the variables or elements, the relationships between the variables or elements, and the bounding of these variables and relationships from the rest of the world. Hall and Fagan's (1956) definition of a system has been used by a number of authors as a starting point (e.g. Thorn, 1988; Huggett, 1980). Their definition is: 'A system is a set of objects together with relationships between the objects and between their attributes' (Hall and Fagan, 1956, p. 18). The importance of the previous chapters for understanding the limitations of systems thinking should be clearer now. The definition of variables and relationships implies an ability to define and divide the world into distinct entities and relations. Likewise, the definition and bounding or closure of the system itself requires a particular view of reality as divisible and understandable by this division.

Systems thinking depends on the reality of the physical environment being displaced from the observer. The observer defines a system made of real entities and relations, the sum of which becomes a sort of super entity with its own properties and relations to the rest of the physical environment. This super entity may or may not be simply the sum of its elements and relations. The observer is a passive and objective interpreter of the system, almost by definition outside of the boundaries they have imposed.

A distinction is often made between a system and a model (Thorn, 1988). A system is viewed as an abstraction that is assumed to exist in reality. A model is a fully specified, although abstract and incomplete, version of reality. A model is an abstraction and a simplification of reality and it is recognized that it does not, nor is intended to, mirror reality. The distinction is made to clarify the purpose-led construction of models as opposed to the supposed universal nature of systems. A model is designed to serve a purpose.

It does not need to specify reality fully, nor to be agreed by all. A system may be unknowable in full, but agreement can be reached that such a set of entities and relationships exists. The distinction may be helpful in distinguishing an operational model from an abstract, but universal, system, but it has the side effect of implying that the system is in some manner more 'real' than the model.

The system has some universal status whilst the model does not. Systems, therefore, seem to be a closer representation of reality despite the user-defined nature of all their entities and relationships. Within physical geography it is usually system models (Thorn, 1988) that are being considered rather than systems as universal abstractions.

APPLICATION OF SYSTEMS THINKING

Development of typologies of systems application has resulted in long lists of different types of systems based on a range of criteria. Chorley and Kennedy (1971) provide one of the first based on the form and function and complexity of the systems studied. Complexity increases from simple cascading and morphological systems to process– response systems and on to biological and social systems. Strahler (1980) develops a typology based on similar criteria to Chorley and Kennedy. Terjung (1976) uses four criteria to separate system modelling in physical geography into different levels. The criteria used relate to the type of logical argument used in explanation (induction or deduction), the level of explanation (individual entities or the system as a whole), the degree of deterministic behaviour and finally the level of description as opposed to explanation. An important basis for these typologies is the increasing openness of the systems. Isolated systems are an ideal type, ones where there is no movement of matter or energy across system boundaries. Closed systems permit the flow of energy across boundaries, but not matter.

Open systems permit the flow of both energy and matter across their boundaries. These distinctions are important as they begin to define the expected behaviour of systems. The definitions derive from physics, where an isolated system will tend towards an equalization of the distribution of energy within it and hence eventually exhibit maximum entropy and disorder or randomness in the organization of its components. Systems that have their boundaries open to flows – particularly if these are of both energy and matter – are able to stave off this 'entropy death' as they retain their organizational structure. Open systems are viewed as able to retain both entities and relationships by maintaining gradients of energy levels between different system components.

Flows from high to low energy are maintained and so the entities and relationships, the network that produces the system, are maintained. Energy and matter are derived from beyond the boundaries of the system to maintain that system. Although the overall entropy within the universe may be heading towards a maximum level, the smaller system being studied is able to reverse this trend within it by importing energy and matter to maintain its order. Entropy is kept at bay by continually exporting disorder and 'borrowing' energy to keep order.

Each of the above classifications implies an increasingly structurally complex view of the world using systems. Within each classification scheme, the simplest level is viewed as description and definition. From this level the structure of the system is identified (the morphology of the system) to which flows of matter and energy are added (the cascading system). Resultant from these two levels is the process–response system, where energy and matter flow through a set of entities arranged in a specific manner. The flows interact with the entities to produce change or stability in these entities and their relationships to each other.

What complexity means, however, is rarely made explicit. There seems to be an implicit acceptance that the lowest level in the hierarchy represents 'mere' description. At this level it seems to be assumed that there is no real explanation. Setting up the system, defining the entities and relationships to be modelled, seems to be viewed as a relatively simple and uncomplex task. This is the level upon which all other levels depend and which forms the basis for 'real' explanation. Complexity seems to be associated with increasing refinement in the specification of entities and relationships. Likewise, a more complex system is one where the entities and relationships are dynamic in the sense of being specified and identified and measured as they change. Full understanding of the system implies that explanation can be generalized and applied to all systems with the same entities and relationships. The problem that full specification would define the uniqueness of the system, and so the uniqueness of its explanation, is not considered as an issue.

One of the most important impacts of systems thinking in physical geography has been the framework it provides for directing and organizing thinking about reality. Thinking is not directed at an individual entity but at the relationship of that entity to other entities and the context of the entity within a system. Not only is thinking directed, but it can also be represented in a formal manner. Representing reality by symbols that can be applied to different systems provides a strong unifying bond between different parts of the subject. This gives the impression of a uniformity of approach and purpose that, although it may be illusory, provides a myth of commonality that is lacking in human geography. It also could be interpreted as a singularity of approach and therefore misidentified as a lack of conceptual and philosophical breadth and depth by human geographers. Symbolic representation of diverse systems also means that they can be modelled, conceptually and mathematically, and so general relationships established and tested between systems. In this manner trends and patterns identified in one part of reality can quickly be transferred and assessed in another part of reality. Systems analysis, from this viewpoint, has helped to identify the holistic and universal nature of physical reality more rapidly and accurately than any technique before it.

Systems thinking and application in physical geography has not been without problems, however. Some of the most important issues relate to the points raised in previous chapters concerning reality and how it is viewed through different philosophies. Central to the problems of systems thinking is the potential to believe that the system is reality. A system is only a model of reality; it is a simplified representation of what the researcher believes to be real and important for the operation of the particular small area of reality they are concerned with. Any system, its components, its relationships and behaviour can only be identified and understood in the context of the theory or theories that have informed its construction. Identifying and maintaining the link between theory and the system constructed should be at the forefront of the mind of any researcher. Without this link the rationale for system construction, and the expectations for system behaviour, is unknowable. Assuming that the system is reality could result in the same system framework being applied inappropriately to different parts of reality.

Systems and Change

Within the application of systems thinking in physical geography an important property that has been developed is the ability to identify and predict certain types of behaviour. In particular the ability to identify when and why systems change or remain stable has become a focus of study. Within systems analysis the types of changes expected are related to how the system and its components change as the inputs change. As inputs, usually considered as discrete events, change in their magnitude or frequency, or both, it is expected that the output from the system or even the internal organization of the system will alter. The type of change will depend upon how the system is organized and the relationships between variables. Importantly, the presence of feedbacks within the system will determine system behaviour. Negative feedbacks will tend to dampen the impact or tendency to initiate change in system behaviour.

Positive feedbacks will tend to enhance the changes caused by any input in system behaviour. Implicit within most assessments of system behaviour is the assumption that the system will behave in a specific and predictable manner. This usually means that the system will exhibit some form of equilibrium behaviour. Equilibrium itself is a complicated and increasingly contentious concept within physical geography, and full discussion of it is left until the next chapter. Chorley and Kennedy (1971) view equilibrium as the maintaining of some kind of balance within a system, whether that be of the relationships amongst the variables, the level of the output or some steadily moving set of conditions. This highlights the idea of equilibrium equating to balance and being the expected or 'normal' behaviour of the system.

The idea and significance of change in the physical environment did not start with systems analysis, but the framework provided by systems thinking has been useful for formalizing the different ideas about change. To clarify how systems analysis has helped to codify change and stability it is useful to review how time has been seen within physical geography and then how change has been incorporated into physical systems. Change (and stability) can occur in both time and space, although it is temporal change that has been the focus of most study. Discussion of temporal change has been seen as a key ingredient in defining contemporary physical geography. Schumm and Lichty's (1965) paper on 'Time, Space and Causality in Geomorphology' is still viewed as a classic presentation of different scales of temporal change, even if the theory is untestable.

They identified three types of time: cyclical, graded and steady-state. At each temporal scale different variables will be important for the operation of the physical system. As the scale changes what were previously important variables effectively become constant, unchanging variables, the context of the system. Although Schumm and Lichty were aiming their critical review at the Davisian model of landscape development, their use of concepts from other parts of science such as equilibrium in developing their argument provided a theoretical basis for changing the emphasis of what geomorphologists did. They highlighted the need to focus on the appropriate scale of study and the variables appropriate to that scale.

Davis was not disproved, merely regarded as irrelevant for more process-orientated studies which focused on reality at a different scale from Davisian landscape evolution. In this respect, although Schumm and Lichty did not use system terminology, their study highlights the importance of appropriate boundary and entity definition for studying reality. Likewise, the use of diagrams such as the one above provided a set of expected behaviours for the scale of reality being studied. They had started to define what type of changes should be expected in particular variables under certain conditions. They had started to define systems behaviour in geomorphology.

Wolman and Miller (1960) looked at how the magnitude and frequency of events could be used to understand change and stability within the physical environment. Although they limited their definition and exemplification to fluvial systems in temperate environments, their form of analysis did provide an insight into change. They identified the magnitude of an event by how much 'work' it did in a catchment. Work was defined by amount of sediment moved by events of given frequency. Importantly, the sediment moved had to be measured as output from the catchment to count as having been 'worked on' by an event. A 'normal' event was viewed as bankfull discharge. This definition limited work to a relationship with sediment movement out of the catchment. Movement within the catchment, or work done in weathering material for transportation, was not included, for example. With this definition of work it is clear that large events do most work – they move most material.

Wolman and Miller also noted that events of a given magnitude occur with differing frequencies. Large events are relatively rare, small events are common. Combining the two trends for events, they produced the magnitude/ frequency curve. From this curve they identified that medium-size events are doing most of the work within a catchment. They linked these medium-size events to bankfull discharge within the catchment and identified forms such as river banks as being adjusted to these size events. They were, in other words, defining expected behaviour for the catchment system in terms of river bank morphology and 'normal' events. The relation between form and event was, however, defined only in terms of the absolute magnitude and frequency of the event. The form and its reaction and interaction to an event is assumed to be constant, the system is devoid of feedbacks. They did, however, draw a distinction between events that may be formative, that may produce landforms in the landscape, and those that did the 'normal' work. In other words they recognized the potential for qualitatively different types of events in the landscape.

Wolman and Gerson (1978) developed a more complicated version of the relationship. In this paper they define event effectiveness – not in absolute terms but in relation to the ability of the landscape to 'restore' itself. This assumed a tendency for the landscape as a whole to move towards a preferred characteristic or equilibrium state once a disrupting event is removed. The event is still viewed as a discontinuity to the system, as an interruption to the 'proper', normal behaviour of the system. Once the event disappears, the system, in their case the landscape, can once again return to its normal modes of operation. Events are not seen as important parts of how a system operates in these papers, instead they are seen as disruptions to be endured.

Event effectiveness did not solely depend upon the property of magnitude. The context of occurrence was vital. If two large events occur one after the other, the second event would have no sediment to remove from the catchment as the first event would have already removed it. Event sequences became important, so the event in the context of other events had to be considered. In addition, an event of a given magnitude would have a differential impact depending on the power of the restorative forces in a catchment. Although Wolman and Gerson tended to limit these forces to vegetative regrowth, the concept could be applied more widely. This new definition highlights that the context within which an event occurs is important for the relationship between event magnitude and frequency and landform change. In other words, the behaviour of the system depends upon the system as much as it does upon the input into the system.

The presence of strong negative feedbacks can dampen the changes initiated by an event. Weak negative feedbacks reduce the ability of a system to 'repair' the damage of an event. It might even mean that the system could not return to its previous 'normal' mode of behaviour. Brunsden and Thornes (1979) added a further dimension to expected system behaviour in their version of landscape sensitivity. They defined landscape stability as a function of the temporal and spatial distributions of the resisting and disturbing forces within a landscape. This concept can be expressed by the ratio of the magnitude of barriers to change and the magnitude of the disturbing forces. The sensitivity of landforms to internally and externally generated changes can be expressed through the transient-form ratio:

TF = mean relaxation time /mean recurrence time of events

This definition highlights the significance of both temporal and spatial changes in a system due to an event. Events are no longer simple single inputs, their nature can change and in changing their nature the behaviour that they induce changes. Likewise, systems are no longer viewed as simply responding to an event, the event is instead mediated through the system and the mediation itself becomes a form of creeping change within the system. Events still have impacts, but the impact is more dispersed and complicated than previously envisaged. System behaviour still adheres to the idea of equilibrium, but the nature of change depends more upon the interaction between system state and input rather than just input. In Brunsden and Thornes different time-scales for individual landforms become important. Time can be divided into the time taken to recover from a disruption (relaxation time) and the time period during which the 'normal' or characteristic form is present.

The characteristic form is the form the landform takes when it is able to absorb disruptive events. Relaxation time is a time period when a landform is adjusting to a disruption and when it exhibits non-normal forms. These transient forms are not in equilibrium with the processes forming the landform; in other words, they are not characteristic forms. Landforms will have different stabilities and react to the same disruptions in different ways depending on if they are adjusting or adjusted to disruptions. Figure 6.4 illustrates this idea by viewing the landform as being entrenched to differing degrees in its current state. Different parts of the landscape will have different degrees of entrenchment and so different sensitivities to the same event.

Barrier to Change

In this way Brunsden and Thornes provide a very complex and spatially and temporally differentiated view of landscape and landform behaviour, but retaining at heart the idea of a tendency to some equilibrium condition. From the above the significant aspect for the behaviour of the system is the relationship between the resisting and the disrupting forces. These general concepts can only be made operational by the definition of the researchers. There are no hard and fast rules as to how to define each of these properties, but their conceptualization is vital for understanding how systems behave. Brunsden and Thornes's analysis requires that entities in the system be defined in relation to an input or event as either resisting or enabling its mediation as a disruption to the 'normal' operation or behaviour of the system. In this manner, the definition of normality or expectations of system behaviour become essential to defining what to study and how to understand changes in system behaviour.

Understanding the system is inherent within the manner in which the system is constructed. Schumm (1979), in the same volume as Brunsden and Thornes, further developed the concept of thresholds in a system and the potentially complex behaviour that even simple thresholds could produce. Thresholds make points in the system where the behaviour changes from one mode of operation to another. In other words, the system stops following one expected course of behaviour and, after some brief odd behaviour, settles down into a different but expected (i.e. predictable) course of behaviour. Thresholds can be extrinsic, a system responding to an external influence. Thresholds can also be intrinsic, with the switch in behaviour occurring without any change in the value of the variables external to the system. Adjustment of a landslide to the continual process of weathering of its regolith could be viewed as an internal threshold. At some point the continual operation of weathering agents could weaken the strength of the regolith sufficiently to induce a failure during a rainfall event that previously would have passed without incident. Schumm identifies a geomorphic threshold as one that is inherent within the system, one that develops through changes in the morphology of the landform itself.

A system does not necessarily respond rapidly to a threshold being crossed. Schumm uses the example of an experimental channel where there has been a single change in base level. Incision at the river mouth progresses upstream, rejuvenating tributaries and removing previously deposited sediment. As this erosive wave progresses upstream, the sediment load increases in the main channel with consequent deposition and aggradation in the previously incising channel. Eventually, the tributaries become adjusted to the change in base level and the sediment supply to the main channel dries up. This initiates a new phase of channel incision. From a single event, the change in base level, a series of incised channels and terraces have been formed and different parts of the catchment have responded at different rates and in different ways to the same stimulus. This experiment, plus the example of Douglas Creek that Schumm uses, implies that even a simple system can have a complex response to events. The precise sequence of changes will depend on the context of the system and its thresholds. Importantly, there remains the problem of identifying sensitive parts of a system and the thresholds before changes occur.

Systems and Landscape Sensitivity

Landscape sensitivity as expounded by Brunsden and Thornes (1979) is an important means of understanding systems and change. In explaining the idea, and in developing its application (Brunsden, 1990, 2001), use is made of major concepts in systems thinking. The application of these concepts provides a systems framework that could be applied to most phenomena in the physical environment. From a Popperian perspective this could call into question the basis of landscape sensitivity as a theory, casting it instead as a myth. In Brunsden and Thornes (1979), four fundamental propositions concerning landscape development are put forward as the basis for constructing the idea of landscape sensitivity. These propositions could be viewed as basic postulates of the theory, or as untestable truisms in the sense put forward by Haines-Young and Petch (1986).

Brunsden and Thornes (1979) view landforms as either being in equilibrium with environmental conditions (characteristic forms) or as forms moving towards that state (transient). The landscape is continually in a state of change as external inputs force it to adjust towards its characteristic state or as internal thresholds are crossed. The distribution of disruptions, and therefore of landforms in different states of transience or stability, is both spatially and temporally complex. The result is a complex assemblage of landforms in varying stages along a sequence of adjustment. The susceptibility of any specific part of a landscape can be expressed by the relationship between the forces resisting change and those forcing it. In other words, there are spatial and temporal variations in the balance of the landscape and its components. Sensitivity expresses how close the landscape and its components are to the edge of this balance.

Proposition 1: For any given set of environmental conditions, through the operation of a constant set of processes, there will be a tendency over time to produce a set of characteristic forms.

Proposition 2: Geomorphic systems are continually subject to perturbations which may arise from changes in the environmental conditions of the system or from structural instabilities within. These may or may not lead to a marked unsteadiness or transient behaviour of the system over a period of 102–105 years.

Proposition 3: The response to perturbing displacement away from equilibrium is likely to be temporally and spatially complex and may lead to a considerable diversity of landforms.

Proposition 4: Landscape stability is a function of the temporal and spatial distributions of resisting and disturbing forces and may be described by the landscape change safety factor, here considered to be the ratio of the magnitude of barriers to change to the magnitude of the disturbing forces.

Brunsden (2001) emphasizes that landscape sensitivity permits an assessment of the likelihood that a given change in the controls of a system will produce a sensible, recognizable, sustained but complex response. The image of the landscape system is one that is in a state of possible change. Whether change occurs or not depends upon the characteristics of the system, but always in relation to the disruptive forces affecting it. The landscape system is balanced between its propensity for change and its absorption of the disruptions within its existing structure. Brunsden (1993, 2001) outlines several sources of resistance to or absorption of disruptions.

These usually reflect the structure or morphology of the system under investigation. System response will also depend on the sequence of events experienced. Different sequences will activate different pathways in the system and so create different responses. As the system responds to events it alters its own characteristics, and so what Brunsden calls 'time-dependent preparation processes' can alter the sensitivity of the system. Weathering, for example, can weaken the bonds within a rock mass and increase its susceptibility to failure by future disruptive, triggering events. Brunsden refers to Crozier et al.'s (1990) model of slope-ripening by increasing regolith thickness through weathering as an example of a preparatory process. Operating in the opposite direction, changes in the system can increase its resistance to disruptions. Church et al. (1988), for example, suggested that the coarsening of bed surface textures occurs in tandem with a rearrangement of the coarse clasts into increasingly stable geometric arrangements.

System Properties Description

Strength resistance Barrier to change imparted by the properties and dispositions of the materials out of which the system is made. Morphological resistance Variable distribution of potential energy across the system. Scale dependent depending upon the processes considered. Structural resistance Design of the system – components, topology, links, thresholds and controls. Subdivisions are location resistance (location of system elements relative to that of the processes capable of generating change), transmission resistance (ability of system to transmit impulse of change). Filter resistance System control and removal of energy from landscape – shock absorbers of the system. System state resistance Ability of system to resist change because of its history. Past history will have configured system pathways in a specific, unique manner, so no two systems will respond the same to the same input.

Development of landscape sensitivity has resulted in a vast number of related ideas that emerge as logical extensions of the initial propositions. The couching of each in systems terminology provides a means of incorporating these into systems thinking. The range of ideas, however, is almost too inclusive. Little in terms of change or stability does not come under the descriptive preview of one of these ideas. Testing the presence of a specific type of change is difficult. In the case of preparatory process, they can only be identified after the change; before the change they are only one of many possible processes of change. Similarly, resistances are defined by the nature of the system, an entity identified and populated with elements by the observer. System resistance is built by the knowledge and experience of the observer, but it is unknowable if there is a basis or even correspondence for this in objective reality.

SUMMARY

Systems analysis has been a dominant method of thinking about the physical environment since at least the late 1960s. Systems analysis provides a set of formal and standard terms for translating the physical environment, a common framework for analysing the whole of the physical environment. Systems are still, however, a simplification of reality, not reality as it really is. Despite some discussion of the meaning of the term, systems thinking has been applied extensively. Systems thinking has, however, constrained the type of behaviour researchers expect from the physical environment. Specifically, features such as positive and negative feedbacks and equilibrium are assumed to be present and to be 'explanations' for behaviour. Likewise, there has been a tendency to use and apply system terms, such as 'robustness', 'sensitivity' and 'relaxation' to the physical environment. These terms have very flexible definitions, depending upon the context in which they are applied. This makes the testing and falsifying of theories expressible in these terms problematic as many translations of these terms are possible.

PARADIGMS AND SOCIAL NETWORKS

The idea that scientific thought and its change is strongly influenced, if not determined, by the society within which it exists is something that Kuhn developed into the concept of paradigms. Although a paradigm is not a well-defined concept (at least 54 definitions according to Masterman, 1970) it has acted as a useful beacon for studies that emphasize the importance of social context for science and, probably more contentiously, for the idea that changes in scientific ideas are illogical and totally socially constructed. Even if you do not subscribe to the very strong social or even socially determined view of change in scientific ideas, the development of science within a social vacuum cannot be sustained any longer. Science, and by implication physical geography, develops within a range of social networks which, even if they do not impact upon the logic of selecting theories or paradigms, certainly do contribute to what and how reality is studied.

Physical geographers are part of society; in fact, they are part of a number of interlocking and interweaving societies. This detail is often lacking in a discussion of both paradigms and of the social construction of reality. The assumption of a single and easy to-identify 'society' with, likewise, simple and singular 'social influences' is a fallacy that requires some correction. It may be more appropriate to think of any physical geographer (or indeed any scientist) as located within different social networks, different sets of social relationships, that can influence how they perceive, study and hence understand reality. Physical geographers are members of humanity and as such are constrained in their behaviour and thoughts. This general level is, however, not usually the sense in which the term 'society' is used in papers exploring paradigms and the social construction of knowledge. A more common use of the term is in relation to membership of either Western or capitalist society. At this scale, the scientist is often viewed as some sort of automaton playing out, more often than not in a very negative and detrimental manner, a set of capitalistic and imperialist imperatives.

This form of analysis has echoes of the simplistic analysis of imperial imperatives used in studying geography in the nineteenth century. Patterns of behaviour and study may be recognizable, but the subsuming of every physical geographer in the UK or USA as a direct and unthinking agent of capitalistic dogma, and every action and thought as being explicable in such terms, is highly reductionist and overly simplistic. Even if actions and thoughts can be interpreted as influenced by the individual being in a capitalist society, the other social networks of which the individual is a part, as well as the individual her/himself, provide the framework within which such a broad social influence is mediated. It should not be expected, therefore, that classifying individuals as from the same society at this scale confers some sort of explanatory power onto that level. Any influence is mediated and negotiated through different social networks. As with imperialist studies of geography, this can mean that different individuals can have very different ideas about what the imperatives of their societies are.

Moving down from the scale of capitalistic society, the problem of social influence upon scientific thought becomes more problematic and more interesting. An individual physical geographer is bound into social networks at different levels. The individual is a member of social networks defined by space, such as a nation or even a nation-state, as well as more regional and local communities. At the smallest scale the individual has a tight and close social network of immediate work colleagues, friends and kin. Although often only the 'professional' relations are highlighted in an analysis of social influences, the other relations do not disappear just because the researcher focuses on the professional.

Any individual will be influenced by social networks other than professional ones, if only in the amount of time they devote to these other networks. An individual will also be located within seemingly aspatial social networks such as class, economic and international networks of fellow professionals. These seem aspatial as they do not require space for their definition, although they do require a physical, and therefore spatial, manifestation for their operation. As the scale at which social networks are considered becomes increasingly refined, the detail required to understand the networks increases and the potential range of influences increases. The individual will respond and act within these different networks in different and not necessarily consistent ways. This makes the unravelling of a model of general social influences extremely difficult. The uniqueness of the individual and their interactions with their individually constructed social networks make such generalizations difficult and often irrelevant to understanding how ideas develop and change.

The above is not an argument for neglecting the social influences on the development of ideas in physical geography. Instead it is a plea for a more subtle and thoughtful approach to what the nature of this influence might be. The individual physical geographer should be placed in their complete and complex social networks rather than caricatured as unthinking and unfeeling passive agents of social trends. It is as responsive and emotive individuals that physical geographers interact within their different social networks, and indeed with different physical environments, and from this interaction that their beliefs and opinions concerning concepts in physical geography develop. Despite the above comments, some general models of the influence of social networks on physical geographers and the development and selection of ideas have been developed. Stoddart (1981) provided a very general model of the hierarchical nature of the social networks within which physical geographers are embedded.

The important aspect of the model is how the individual mediates the social influences. Not all social influences are the same, however. Some social practices may not be viewed as such as they have become ingrained within the social networks as expected norms or rules of behaviour. This is equivalent to the 'disciplinary matrix' identified by Kuhn. It is the tangibles, such as textbooks and written documentation, and the intangibles, expected norms of behaviour, that make up the professional practice of 'doing' physical geography. The rules defined by such practices could refer to the type of questions to be asked, the methods used to answer particular questions and the forms of analysis employed. These practices are not questioned, and once established by continual practice and repeated publication they become viewed as the way to undertake a particular type of study or the route to understand a particular phenomenon.

9

SOCIAL CONSTRUCTION AND PHYSICAL GEOGRAPHY

Recent discussions in geography have focused on the social construction of reality. Papers by Demeritt (1996, 2001a) in relation to climatology have put forward the argument that the reality studied by physical geographers is socially constructed. Countering this view, Schneider (2001) has suggested that such a strong social constructive view is detrimental to the subject. Specifically, this view highlights the relative nature of knowledge and denies the possibility of objective knowledge of the physical environment. Whilst social construction of reality is undeniable in the sense that physical geographers always work within social contexts, the strong 'sociology of science' view, that particular brands of social construction demand, paints a relativistic picture of knowledge that most, if not all, physical geographers feel uncomfortable with.

Discussion of social constructionism should begin, as Hacking (1999) notes, with assessing why you need to assert that something is socially constructed. Social construction in relation to social phenomena usually develops an argument concerning the negative nature of the phenomenon under investigation. Central to the argument is the seemingly natural nature of the phenomenon despite its social construction. There is no discussion of the reality of the phenomenon; its negative nature is to be negated by its social construction and, by implication, its non-existence in an objective reality. Applying the same argument to physical geography implies that a physical and real basis to the entities and properties studied does not exist. This view of reality is at odds with other philosophies such as critical rationalism, critical realism and pragmatism which all assume the existence of an independent reality. Although these all also do not assume that this reality can be known directly, they do assume that it exists and that physical geography is striving to interpret it. Social constructivism, in its strongest sense, appears to deny this possibility or even encourage any attempt towards it.

It may be that the subject matter that is the focus of physical geography is different in nature from the subject matter of human geography. This is not to say that social constructivism, in the 'weak' sense identified above, cannot be applied to the subject matter but rather that the presence of an underlying reality that the subject matter refers to is not denied. As noted in Chapter 3, Hacking identified two types of kinds that can be the focus of study. Indifferent kinds are kinds or entities that do not react to the process of being classified. Interactive kinds are kinds or entities that do react and can respond to the process of being classified. Physical geography tends to deal with indifferent kinds or entities. They reflect the dialogue between the researcher and an independent reality.

The indifferent kinds do not necessarily accurately mirror reality, but they do arise from a social constructed, yet consistent, dialogue. It is this consistency and its agreed nature that provides the products of the dialogue with their objective power. Objective in this sense does not mean that the entities or kinds are 'real', only that they have been derived by a consistent process of dialogue and are treated as if real. The exact nature of the entity or kind may alter as the dialogue changes, but once defined the entities or kinds themselves influence the dialogue. In this sense it is pointless to begin to contrast 'real' as opposed to 'socially' constructed phenomena. The dichotomy is a false one. In effect, phenomena are both real and social at the same time.

Ethics in Physical Geography: Reflection Required

Ethical considerations in physical geography have not received much attention within the printed literature, although some debates have arisen at conferences. Scientific study in fields such as medicine and sociology have developed increasingly sophisticated means of assessing the ethical contents of their subject matter. Given that the focus of their research is on humans, this may seem an inappropriate place to look for guidelines about how to tackle ethical issues in physical geography. The basis that these fields have established does, however, provide information on the type of questions that should be considered. Within physical geography there are specific areas that should be questioned. First, at an abstract level the acceptance of an external, independent reality imposes a series of moral obligations on the scientific researcher. If the researcher is committed to uncovering, as far as they can, the operation of this reality, then there is an implicit obligation to 'truth'.

This is not as simple as it appears. Most physical geographers might assume that this means that the researchers should follow procedures diligently and record and revel in success and failure both the same. In other words, the researcher's work should reflect reality: the researcher holds a mirror up to reality and records the reflection without distortion. Hopefully, all of what has gone before illustrates the impossibility of such a naïve view of reality and its scientific study. Commitment to truth could be interpreted as ensuring that you derive data that reflect how reality is. Your view will be driven by theory and so truth, and these will become one and the same thing to you. Denying the theory is to deny the truth of how reality is. Reality may be interpreted to fit your theory. Ethically the researcher is committed to the truth, but other researchers may deny their version of the truth.

Commitment to the truth could be interpreted in another way. It could be viewed as commitment to use a specific approach to understanding reality and to abide by the outcomes of this approach. The centrality of falsification and testability in the philosophies outlined previously provides another means of reflecting reality. Researchers admit to the fallibility of their views of the world. Researchers submit their theories to rigorous testing in a dialogue with reality and with their co-researchers.

Commitment to the truth then becomes commitment to this process of scientific scrutiny. It is through this method of rejection of theory that reality is to become knowable. Adherence to this process does, however, rely on an agreement between the researchers that the process of scrutiny is 'fair' and legitimate. Mechanisms for legitimating the process have evolved with the subject, from initial acceptance of the accounts of 'gentlemen' to the complicated peer-review systems for academic papers and research projects. The development of legitimate pathways does, however, also constrain research that is acceptable or capable of judgement by the formats made available by these pathways. Second, physical geography is embedded within social networks as previously noted. These networks include moral ones that are variable in space and time. These networks may dictate what is an appropriate set of questions to ask, and what are appropriate methods of scientific investigation and appropriate forms of explanation. The early history of the study of the physical environment is littered with what seem now to be irrelevant moral constraints. Chorley et al. (1969) illustrated the need to retain the study of landscape development within a temporal framework that respected biblical events in the seventeenth and eighteenth centuries.

The development of a catastrophic view of landscape development then produced a reaction of strict uniformitarianism within geological and evolutionary thinking. Within contemporary society, of which physical geographers are a part, there is increasing discussion of the moral rights of the environment and its components. Issues of preservation and conservation of the physical environment impinge directly on the subject matter studied by physical geographers.

The indifferent entities and kinds that physical geographers study are part of the physical environment. Rivers, rocks and soils all contribute to the landscape, which has become a focus for moral concerns both in its own right and as the basis for supporting ecosystems and particular rare species. Concern that these entities are preserved for future generations as they are now, or conserved as dynamic entities, is the basis for this moral concern. A whole literature has developed concerning the moral and legal rights of future generations. Likewise, this concern has resulted in the development of sets of principles by which research activity is judged in terms of its impact upon the physical environment. Principles such as minimum impact and minimum tools are used within the US National Park Service to guide the assessment of the environmental impact of scientific research. These principles define the type of practical steps researchers need to build into their experimental designs to be granted permission to work within national parks. Principles such as minimum impact, for example, ask the researcher to consider how they reach the site. Do they take a car, what damage do they cause by their interference? and so on. The minimum-tool principle likewise requires the researcher to justify data collection using the proposed method. Issues such as environmental damage using a particular instrument relative to other instruments are put near the top of the agenda.

Each principle and its practical implications are designed with resource conservation in mind. The principles begin to question the assumptions researchers make about what are appropriate methods of study and bring to bear issues of general responsibility for actions that may have rarely entered into a researcher's consideration before. Third, a more general view could be taken of the ethical considerations within physical geography. The practice of physical geography involves a range of ethical choices, some of which are not viewed as choices at all but as established working practices. Recent discussions of the position of women within the earth sciences, for example, raise the issue of whether the practice of physical geography is inherently biased. The number of women, and the positions they occupy, could be viewed as being an indicator of gender-bias within physical geography. The development of an academic community requires that individuals acquire certain qualifications and become members of certain groups to be accepted in the field. Is selection of candidates for Ph.D.s based on gender lines?

Do decisions about the employment of academic staff get caught up with expectations of social worth and individuals' expected life choices (i.e. women get pregnant and look after the children, so are to be expected at work less)? You would hope to be able to say that in contemporary society these issues would not be a concern and that the answers to the above questions would all be an emphatic no! Suggestions such as physical fieldwork is based on a macho image of a fieldworker, and that the society to which they belong (e.g. a drinking and male-orientated culture) sometimes make it difficult to answer the above questions confidently. That these issues are rarely aired in physical geography does not mean they are not important.

In a similar vein there has recently been concern for the 'invisible' researchers within geography. The Research Assessment Exercise in the UK has thrown into sharp relief the role and lack of career structure for post-doctorate researchers and even Ph.D. students. The trends in the employment of this group could be viewed in the same light as the general trend towards casualization of the labour force as a whole with all the associated problems. Academics, often the employers of these individuals, have been relatively slow to appreciate and develop strategies for career development. Often academics working with short-term contract researchers do not see it as part of their job to think about their employees. Often the employees are unaware of what their rights are. Likewise, comments are usually made about the increased flexibility of work practices for researchers by full-time academics.

The dialogue between the groups, if indeed there are identifiable groupings, is one of an unequal power relationship. Oddly, academics have rarely researched this personal and close relationship between themselves and their researchers. Increasingly, in the UK at least, as research funds become concentrated in a few 'top' institutions, the development of academic careers will be controlled by fewer and fewer individuals. The ethical implications of this have rarely if ever been discussed. The above discussions illustrate the important role that issues of responsibility play in developing an ethical code for research in physical geography. Although policy instruments such as minimum impact and minimum tools impose a framework of compliance upon researchers, they can also be viewed as practical and codified practices for environmental responsibility. The development of an ethical framework can be imposed from above, as in this case, but it is more relevant to ask why the framework has to be imposed in the first place. Questions of moral uncertainty may be part of the answer, but it is likely that unaddressed issues of researcher responsibility are also part of the problem. Scientific research tends to focus on a particular problem.

As the book has highlighted, and as others have commented, science is the art of the possible. Researchers break questions down into manageable chunks and answer what they believe they can using the tools that they have learnt to use. Questioning of the questions being asked and the impact of their investigations upon the state of the physical environment for future generations is not a question directly involved in their framing of research. Particularly where research is focused around the short-term research grant with its associated promotional rewards, individual justification for an action may outweigh potential ethical concerns if they are even articulated.

Development of an ethical framework needs to be developed within the individual researcher. A researcher may claim they have an ethical commitment to discover the 'truth' of reality, but what do they do if the method they intend to employ causes irreversible damage to the environment they are studying? Which ethical concern has the greater moral imperative? Development of such an individual ethical framework will require a lot more work at the level of individuals and their training, as well as being translated as a payback in career success. If the context of the individual researcher is not given as much weight as the ethical argument, then any ethical framework will founder on the grim personal practicalities of the individual.

Although great consideration has been given to the ethics of research in the physical environment, the same consideration should be given to ethics in the process of research. Increasingly, research is a group or team activity rather than the solitary undertakings of a lone scholar. Similarly, with the advent of the Research Assessment Exercise in the UK, as well as the traditional peer-review system for publications, success of research activity is dependent upon a community. How individuals deal with other individuals in this research process is as much the sphere of ethical considerations as is the output from the process, the research itself.

Once again this sort of issue is something that academics have not had to deal with and often it does not even register as part of their ethical framework. Within the UK, for example, there has recently been a concern over the casualization of researchers, almost the development of a research underclass. The status of these individuals in the research process is often left to the vagaries of individual research team leaders.

The responsibility of employer to employee and vice versa are part of this relationship, but one which academics are rarely trained for or necessarily have an ethical framework in which to assess. It is telling that a number of universities have developed rules for inclusion and precedence of authorship on academic papers involving such teams. These raise issues of unequal power relationships within the research process. The inequality of these relationships requires a recognition and consideration that has been lacking in the recent past.

PHYSICAL AND HUMAN GEOGRAPHY: DIVISION OR INTEGRATION

Recent discussions over the nature of geography often refer to the human/physical divide in the subject. Despite protests and pleading to the contrary, the divide is often seen as insurmountable, with some even claiming that geography has become effectively split into two specialisms (some even carry the subdivision amongst specialisms further). Massey (1999) tried to tease out the commonalities between the two parts of geography via a consideration of the role of entities. Raper and Livingstone (1995) confirm the potential for discussion, as did the ensuing printed debate. Despite these overtures across the divide, there seems to be little real movement of positions so far. This debate about the possibility of integrating physical and human geography is not a new one. A special issue of the Transactions of the Institute of British Geographers in 1986 provided a series of short papers on different geographers' opinions on the likelihood of such common study. These concerns and hopes have also been expressed further back in the century, as in Man's Role in Changing the Face of the Earth (Thomas, 1956). Several geographers have explicitly been concerned with developing concepts that would provide this integration, such as Carl Sauer and the concept of landscape (1956). By the end of the twentieth century, the debate still continued without resolution.

Keeping the two sides apart are centrifugal forces that vary from place to place, but which collectively are likely to prevent the integration wished for by many. Academics, as noted above, are only human. Career paths and funding requirements mean that people will pursue courses of action profitable to them in their particular context. If funding councils do not recognize an integrative physical/human geography then why pursue it?

There is no money in it; there are no rewards. This may sound cynical, but an academic life is short, the period available to make an impact even shorter, so why waste it in pursuit of an unattainable goal? There is also the feeling amongst key researchers on both sides that what they study is different from that being studied on the other side. The subject matter of human and physical geography is seen as fundamentally different. Hacking's (1999) distinction between indifferent and interactive kinds may be taken by some to be a good description of the differences between physical and human geography respectively.

This may be simplifying the matter, but it does have an appeal in explaining the continued divide. If the subject matter is different at a fundamental level then the techniques employed to study one kind cannot necessarily be easily transferred to the study of the other kind. This means that a reductionist approach will not be of any use in a post-modern analysis of asylum seekers. Likewise, what could a post-modernist say that is meaningful to the study of gravel-bedded rivers? Acceptance of such a key divide will justify the lack of communication between the camps.

With such entrenched and well-funded camps is there any point in pursuing the issue of integrating physical and human geography? Maybe, unsurprisingly, my reply is yes! Although some may regard any integration as superficial and part of a propaganda drive to maintain a university subject that has run its course, I still believe that there is room for an integrative approach.

The methodologies of human geography can be applied to the study of the practice of physical geography. Collier and Inkpen (2002, 2003), somewhat simplistically, began to analyse how surveying at the RGS in the nineteenth century was influenced in its practice by the context of the period. Livingstone (1984, 1987) has analysed Shaler's work in both eugenics and geology in the context of the development of Darwinian concepts at the time. Such historic studies can help us to see how seemingly objective facts and concepts are actually constructed by individuals in societies and how the nature of 'objectivity' is affected by these contexts.

Likely to be of more significance to current physical geography is the increasing recognition (stating the obvious maybe) that it is difficult to separate human and physical environments. The interpenetration of the physical and human means that it is difficult to justify that processes of environmental change are purely physical or that social structures rely solely upon human processes.

Maintaining that one realm does not influence the other would be to view each as meshing together at increasingly smaller scales, but never actually overlapping or merging. Gould (2001) provides a similar image of the relationship between science and religion, labelling it as 'fractal' as the two spheres mixed, but never merged no matter the scale of resolution they were observed at. For human geography I mean the more humanistic approaches that seem to address different questions and a different reality from the 'scientific' approaches.

The seemingly clear divide between the two becomes messy when closely examined in terms of, say, the purpose of study, the questions asked, the entities and processes identified. The two, however, never mix, their realities remain different but interwoven. The two never merged as each answered questions about a specific realm: science, what reality was like; religion, how we ought to live. One could not answer questions about the other. Keeping human and physical geography separate implies the same key distinction. Interestingly, however, this separatist view would not actually deny the importance of either branch of geography; it would merely accept that human and physical questions were distinct and separate, but equally as valid to ask in the first place. This view would enable an uneasy truce to prevail in conflict over the integration or division of the subject.

An important consideration that has rarely been considered is that both human and physical geography are not homogeneous masses. There are distinct approaches within both with differing methodologies and views of reality. Arguments for separation, and even some for integration, have usually focused upon the extreme members of both realms.

The argument seems to run that if these end members cannot be integrated then none of physical and human geography can. It may be that these end members follow Gould's fractal description, but other parts of human and physical geography may be easier or rather more appropriate to integrate. The approaches and topics that ask the same questions about reality or can inform the questions that each subdiscipline asks may be more appropriate candidates for integration. As environmental degradation continues, this may be an area that of necessity becomes a concern for geography as a subject for both its relevance and survival.

SUMMARY

Physical geographers work within different societies at different scales. All are members of different networks and may behave in different, even contradictory, manners, depending on the network being considered. Individuals should not be seen, however, as automatons responding blindly or consistently to their context. Context is mediated by individuals in a complicated and often unique manner. Recently, ethical concerns have become an important consideration in physical geography, whether forces upon researchers through instruments such as minimum impact or from within by moral considerations. There is little in terms of a coherent and explicit professional code of conduct for physical geographers to follow. Ethical obligations to the truth, to the physical environment being studied, to other researchers and to the individual are all aspects of the research process that are not usually considered in detail.

10

MINERAL PRODUCE AND RESOURCES

INTORDUCTION

There are several valuable coal fields in India, some of which are being successfully worked. The most important of these is in Bengal, south of the Rajmahal Hills, from which point it extends westwards to the valley of the Son, and from thence south-eastwards through Chota Nagpur to Orissa. Another large field is in the valley of the Godavari stretching north-west from the Ghats to beyond Warora, and extending westwards into the Nizam's Dominions. A smaller field is in the basin of the Narbada, south of Jubbulpore. In Assam there are several coal fields in the valley of the Brahmaputra, where coal of a better quality is found. In the Native State of Rewah is another field which is being successfully worked. In Burma there are some promising fields, but little mining has as yet been attempted.

At present by far the largest output of coal is from the Bengal mines. The chief collieries are near Raniganj, and at Jherria and Giridih. In the Central Provinces there are mines at Warora (worked by Government) in the Godavari coal field, and at Mohpani in the Narbada field. In Hyderabad the Singareni mines are very successful and productive. So, also, though on a smaller scale, arc the Umarift mine in Rcwah (which, like the Warora mine is worked by Government) and the Makum mine in Assam. The total output of coal from Indian mines in 1907 was over 11 million tons, of which nearly

10 million tons came from Bengal alone. This was sufficient to supply the chief needs of the country, and to leave a million tons for export to Ceylon and the Straits Settlements. The import of coal from England has fallen during the last twelve years from three quarters of a million to a quarter of a million tons.

Iron

There is an abundance of iron ore in India, and much of it is of excellent quality. Especially rich ores are found round Salem in the Madras Presidency, in the Raipur District of the Central Provinces, and in Orissa. At one time large quantities of iron were produced in India. But the native system of smelting needed an abundant supply of charcoal, and the decay of the industry was due in part to the decline in the fuel supply, owing to the wholesale destruction of forests. Native smelting is still carried on by isolated groups of metal workers, but the quantity of iron produced is very small, and there is now no possibility of a revival of the industry, as imported iron is much cheaper.

Iron can only be profitably worked in India where, along with the ore, coal and some form of limestone are found, the latter being needed as a flux. If either of these has to be brought from a distance the cost of carriage makes the work unremunerative. At Raniganj, in Bengal, these conditions exist, and iron-works have been established there which smelt about 70,000 tons of ore a year. But the great bulk of the iron used in India is imported from Great Britain and Belgium.

Gold

This precious metal has for ages been found in India. Gold commonly occurs embedded in quartz, a hard crystalline rock of the most ancient geological formation. Rivers which cut their way through such rock almost always bring down more or less gold, which is deposited along with the other detritus which forms their alluvium. Often, through the greater weight of the particles of gold and the variation in the speed of the current, the metal is deposited in particular parts of the river bed. All the greater rivers of India cut through archaean rock, and almost all of them have in the past yielded gold. This is particularly the case with the Brahmaputra, the Irrawaddy, the Godavari, and the Himalayan rivers just as they emerge from the hills. Washing the river sands for gold is still a favourite occupation of many of the hill tribes, and small quantities of the metal are thus obtained.

In Mysore, the Wynaad, and Hyderabad, the rich gold-bearing quartz is mined. It is then crushed and washed. The process is costly, and many mines sunk during the last quarter of a century have not found quartz sufficiently rich in gold to pay for working. Large numbers of ancient, disused, mines exist, which were doubtless the main source of the gold for which India was famous many centuries ago. Almost all the gold now produced in India comes from the Kolar mines in the Mysore State, which for some years have yielded about 2,000,000 worth per annum.

Salt

Of all the mineral products of India salt is in many respects the most important. Native salt supplies the needs of the entire empire, save the Provinces of Bengal and Burma, and forms considerably more than two-thirds of the whole amount consumed. Of imported salt more than half comes from England, more than a third from Arabia, and most of the rest from Germany. The Indian supplies are from three main sources : (1) the sea coast, (2) the salt lakes and pits of Rajputana, of which by far the most important is the Sambhar Lake, and (3) the salt hills of the Punjab and the North -Western Frontier Province.

Along the sea coast salt is manufactured from the sea water by solar evaporation. Both in the Madras and Bombay Presidencies the manufacture is carried on partly at Government factories, but chiefly at private ones under Government license. The Madras salt supplies the whole of South India, the east coast including Orissa, and the eastern part of the Central Provinces. The coast factories of Bombay supply the chief needs of Bombay and Sind, and a part of the

Deccan. On the Rann of Cutch the Government own the Pritchard lirine Works which yield a large quantity of excellent salt, most of which goes to the United Provinces, Central India, and the Central Provinces.

The Sambhar Lake—a lake about 20 miles long, lying a little to the north-east of Ajmere—has water so salt that in dry seasons the crystals gather on the surface to a thickness of six or eight inches. Vast quantities (about one-seventh of the whole produce of India) are taken from it without appreciably diminishing the salinity of its water. What the source of its salt may be is a matter of conjecture. The Sambhar Lake, together with brine springs at Pachbhadrii in the Jodhpur State, 180 miles to the south-west, supply Rajputana, and a great part of the United Provinces, Central India, and the Central Provinces.

In the north-west corner of India ranges of hills containing vast quantities of rock salt stretch east and west on both sides of the Indus. The name Salt Range is, however, confined to those on the east side of the river. There are numerous mines both in the Salt Range proper and in the hills of Kohat west of the river, and in the latter district, where the salt crops out. it is quarried like stone. From these mines and quarries a cheap and plentiful supply is obtained for the Punjab, the Northwestern Frontier Province and Kashmir.

Petroleum

There are three valuable petroleum fields in Burma, the produce of which is rapidly increasing. These are, in the order of their yield, the Yenang Haung, the Singu, and the Yenang at fields. The crude oil is obtained in wells and has then to be purified. The yield of petroleum in 1907 was over i so million gallons, which is more than three-fifths of the entire quantity consumed in India. The imports of petroleum in the same year fell to 60 million gallons. Not many years ago three-quarters of the petroleum used in India was imported, now little more than a quarter. The Burma fields thus promise at no very distant date to supply the entire needs of India. A considerable quantity of petroleum is also obtained at Mlkuni in Assam.

Manganese ore is found in Madras, near Vizagapatam, and, in unusual richness and purity, in the Central Provinces and Central India. The amount raised has increased twelve-fold during the last eight years, and in 1906 nearly half a million tons were exported. Manganese is used chiefly in the manufacture of steel.

Mica of excellent quality is found in Bengal, and in smaller quantity in Madras. About 2,000 tons are shipped to Europe yearly. Saltpetre is found in Bengal, as well as in some of the ether northern Provinces, and about £-20,000 worth is exported yearly. Copper also exists in Bengal, but is not as yet worked. Plumbago is found in Travancore. Tin exists abundantly in the southern parts of Tenasserim, but is not worked to any large degree.

Precious Stones

Burma has valuable puby mines. Indeed, of the best stones it has a monopoly, and supplies the world. Mong with the rubies a few sapphires are also found. The Mines are worked by a Company who hold them on lease from the Government. The Company also grants mining license to private individuals who make no return of their finds. The total output is, therefore, difficult to discover. Jade, a green stone greatly valued by the Chinese, is also found in Burma. Valuable diamonds have been found in the past in the Godavari basin, and are occasionally found now. But there is no systematic mining for them.

Wild Animals

The lion is now found only in Kathiawar, and though for many years it has been rigidly preserved, it is almost extinct even there. The tiger is found in most parts of India wherever extensive forests or jungles exist. Tigers abound in the Terai, the jungles of the Sandarbans, and the forests of Central India and the Western Ghats. The panther, or leopard, is still more common. The cheetah, or hunting leopard, is a different species, and is native only to the Deccan. It is trained for hunting the antelope, but though swift and sure in attack has no staying power. Wolves abound still in the open country but shun the forests. The common black bear is met with in the forests of rocky hills, and the Tibetan sun bear is found throughout the whole length of the Himalayas, but only at heights of over 5,000 feet. Hyaenas;ire nunicrous, but only where the wolf is not. Jackals are common everywhere and are useful scavengers.

The wild dog, which hunts in jiacks, is found in the forests of Burma and Assam.

The wild elephant exists chiefly in the forests of Eastern Bengal and Assam; less commonly in those of Burma and the Indian peninsula. Under a special Act of the Indian Legislature the elephant is strictly preserved, and can only be captured by license. Considerable numbers are caught and trained for the Government service and for sale. The elephant is not found at all in the north-west of India, where the climate is too dry and the temperature too variable for him. Several species of rhinoceros haunt the swamps of the Brahmaputra valley, the Sandarbans, and Chittagong; and their horns are valued by certain classes of the natives. The wild hog is common in most Provinces in the vicinity of cultivation, and hunting him is a dangerous and exciting sport. The wild ass roams in herds of 20 to 30 in the deserts of Sind and Cutch, and several species of wild sheep and goats have their home on the Himalayas, the mountains of the north-west, and the Sulaiman range. The ibex, a species of goat, is found on the mountains of the peninsula and Kashmir. Antelope arc fairly common in Central India and in the coast flats of Gujarat and Orissa. Bison are mot with in the hill jungles of South India and Burma, and the wild buffalo in Burma and Assam. Of deep there are many species, the chief being the sambup. Monkeys of many kinds are exceedingly numerous almost everywhere. Rats and mice abound throughout the land, and the bandicoot, the most gigantic member of the tribe, is exceedingly destructive.

Of reptiles, snakes and scorpions of many kinds are everywhere found. The largest Indian snake is the python, which sometimes attains a length of 30 feet. Most of the snakes found in India are harmless, but there are three or four whose bite is deadly, and for who.se poison no effective antidote is known. One of the worst of these, and at the same time one of the commonest, is the cobra dicapello. Numerous poisonous water snakes are also found. The blunt-no.scd crocodile infests swampy rivers and backwaters, and the sharp-nosed species, the ghavail, which preys only on fish, is numerous in some of the greater rivers, especially the Ganges, Brahmaputra, and Mahanadi.

The destruction of life and property by wild animals and snakes is very great, and shows little tendency to decline. The total number of persons killed by wild animals (chiefly tigers and leopards) in 1904 was 2,157, 000 of cattle 88,206. Poisonous snakes are ten times as destructive of human life, but not so fatal to cattle. Deaths from snake bite in 1904 numbered 21,880, and the number of cattle killed by snakes was 10,376. Rewards are given by the Government for the slaughter of poisonous snakes and many species of dangerous wild beasts.

The birds of India are very numerous and beautiful, but are more esteemed for the gaiety of their plumage than for the sweetness of their song. Parrots abound, and many kinds are made household pets, as also is the maina, a species of starling which can be taught to talk. Small winged game exists in great variety, including snipe, partridge, quail, plover, teal, and wild duck. The peacock is found in the forests of the Deccan, Assam and Burma; the pheasant in the Himalayas; and the red jungle-fowl, from which domestic poultry are said to have been derived, is met with in most parts. The common crow is familiar everywhere.

Of birds of prey the vulture and the kite are the best known. They are everywhere useful scavengers. Eagles are numerous in the Himalayas. Several species of falcon are trained by the natives for hawking purposes. Hawks, herons, and kingfishers of many kinds abound, and the last is much sought after for its beautiful plumage.

Fish of many excellent varieties abound in all the rivers and most of the tanks. The best river fish belong to the carp and barbel families. The mahsir, sometimes called the Indian salmon, is found in the hill streams both of the Himalayas and the peninsula, and grows to a great size. The hilsa is a similar fish, though smaller in size, which abounds in the streams of the Gangetic delta. All round the coasts salt water fish is caught in great abundance, and the fisheries are a source of wealth to many thousands of people.

As in every tropical country insect life is abundant in India. Ants of many species are found everywhere, and though destructive, they are of immense use as scavengers, for they quickly remove every particle of decaying animal matter. The most destructive of all insects, and one that has to be constantly guarded against, is the white ant. The mosquito is, the greatest of insect plagues, and where it alioimds renders life almost unendurable. Fierce war is waged against the mosquito now, as one species has been proved to be the chief distributer of malarial poison. Of useful insects the bee, the silk-worm, and the lac insect arc cultivated.

Domestic Animals

Horses and ponies are common in lower Province of India, but in relation to the population are most numerous in the Punjab, the United Provinces, and the Central Provinces. Indian bred horses are not,- however, so good as those imported from Arabia, Persia, and Australia. Pegu ponies liavc long been famous. Asses and mules are most numerous in the Punjab. The Indian Government breeds mules for use in army transport. They are strong and hardy, and especially useful in hilly districts and rough roads, being more sure-footed than the horse. 200. Cattle, i.e., bulls, bullocks, and cows, are everywhere reared and greatly valued. They are almost equally common in every Province, and their number is everywhere found to be roughly proportionate to the population. The Government has of late years paid great attention to the improvement of the breeds of cattle, and their protection from disease, as well as to the provision of fodder. Sheep and goats also are pretty generally distributed, except in Burma, where they are rare. Sheep are most numerous in Madras, and next in the Punjab. Goats are most numerous in the United Provinces, next in the Punjab and Madras.

The elephant is used for state display by the Native Princes, and for heavy transport by the Government. In Burma elephants are trained to work in the timber yards and may be seen hauling and stacking the heavy logs. For sagacity the elephant is hardily equaled even by the dog. The camel, the most useful of all beasts of burden in a hot and dry climate, is commoner than the horse in Sind, and almost as common in the Punjab and the North-Western Frontier Province. Camels are used also to some extent in the United Provinces, but in other parts of India they are hardly known. The buffalo is common throughout India, and, like the bullock, is used for draught purposes both on the road and in the fields. As in every other country, dogs of various breeds are universal. The most notable are the mastiff of the Himalayas, and the polegar hound of South India.

Economic Animal Products

The ordinary animal products which are everywhere used as food, and, as such, are common objects of local production, trade and consumption, need only be mentioned. Such are milk, butter, ghee, eggs, fish, poultry, game, flesh, etc. Flesh, either of beasts or birds, is a less common article of diet in tropical regions than in colder latitudes, and in India a vegetable diet has the added sanction of religion. Of the natives of India few beyond the Muhammadans and the lowest castes of Hindus are habitual flesh eaters. Milk, butter, and ghee, and all forms of dairy produce, are, however, universally used. So also are eggs. Fish, wherever it can be had, is a welcome food to full two-thirds of the people, and in the deltas of the rivers and along the coasts it forms a staple article of diet.

Fish curing is being developed in India as an economic industry, and it is possible that at no very distant date salt fish may become an article of export. Already there is a small quantity exported from Sind. Fish in India is commonly eaten fresh, and until very recently the process of fish curing was hardly known, or was rendered impossible by the prohibitive price of salt. For some years now the curing industry has been fostered by Government on the coasts of Madras and Bombay. Fish curing yards have been opened, and salt for the purpose has been supplied at a nominal figure.

There is a considerable production of wool in India, though of an inferior quality. Much of it is u.sed locally in the manufacture of carpets, and almost all the rest is exported to Great Britain. The value of the raw wool exported in 1907 was 1, 600,000. For the best class of woollen manufacture in India raw wool is imported.

Hides of various kinds are a valuable article of commerce. They are collected in every Province both for the local leather industry and for shipment to Germany, Italy, and Austria. The value of the hides exported in 1907 was over 10,000,000. There are also small exports of horn, bone manure, and bristles.

Silk is the produce of the silk-worm, which is cultivated largely in Bengal and Assam, and to a less extent in several other provinces. It lives only on the leaves of the mulberry tree. Of late years great attention has been given to the culture both of the silk-worm and of the mulberiy, and the silk industry is increasing. A wild silk is obtained in large quantities in Assam, which is known as tusser silk. It is locally manufactured, and silk fabrics are much used by the Assamese. A good deal of the cultivated silk is also manufactured locally. The value of silk exports in 1907 was over ,500,000. Lac is deposited by the lac insect, and is collected by the hill tribes in the Central Provinces and Chota Nagpur. It is the source of shellac, a material used in the manufacture of varnishes and sealing-wax, and of lac-dye. The lac insect lives on many kinds of forest trees, and its artificial culture has been attempted by the Forest Officers in the Central Provinces. The lac industry is a remunerative and a growing one, and the exports of lac and lac-dye in 1907 were over 2,300,000 in value.

CLIMATE OF INDIA

Blanford remarks that we may speak of the climates but not of the climate of India, for "the world itself affords no greater contrast than is to be met with at one and the same time within its limits." It would be a vain task to describe the climate of every part of India, and any attempt to do so is unnecessary. Climate is everywhere the result of certain conditions whose influence is well understood. The presence or absence of these conditions enables us readily to explain and understand all climatic differences.

Temperature

The Tropic of Cancer crosses India almost midway between its northern and southern points. Passing through Cutch on the west and the Gangetic Delta on the east, it very nearly marks the division between peninsular and continental India. The whole of the peninsula is within the tropics, and Cape Comorin is just over 8° north of the equator. The Indo-Gangetic plain, on the other hand, lies outside the tropics, but near enough to be within the region of greatest solar radiation in the summer months.

In the absence of all other modifying causes, therefore, we should expect the south of the peninsula to have the highest mean annual temperature, and the lowest annual range. Passing north we should expect the mean annual temperature steadily to diminish, and the mean annual range to increase; while from 2° south of the Tropic to 6° or 8° north of it we should expect to find the summer maxima higher than in any other part of India. In the main this is the case, but in India, as in all other countries, the presence or absence of water, the prevailing winds, the proximity of mountain chains, elevation and aspect often make the ordinary temperature of places in the same latitude totally different. It is easy to see how very greatly the climate of India is moderated by these various causes. Water in all its forms is the great moderator of heat and cold. Happily the greater part of the country is, as a rule, sufficiently well supplied with water to render extreme day and night temperatures impossible, and in many parts the climate is remarkably equable. The influence of the sea is also felt far inland all round the peninsula. In Rajputana, Sind and Baluchistan, however, the daily range is often so great as to be extremely trying to all but the most robust. In the highlands of Baluchistan during the late summer or early autumn a day temperature of 80° F. is often followed by a night minimum of 10° F. The air is exceedingly dry, the ground a mixture of rock and sand, and radiation proceeds with amazing rapidity.

It is in this connection that the nature of the soil, and the presence or absence of vegetation, exercise a powerful influence on climate. Some soils are shallow and porous and rest upon a bed of impenetrable rock, so that the rainfall quickly flrws away and they are soon dry. Such is the character of the greater part of the east and south of peninsular India. Other soils have a remarkable power of absorbing and retaining the rain that falls upon them.

The black "cotton soil " that prevails over the greater part of the north-west of the peninsula and Kathiawar, and the mixture of clay and loam which forms the eastern part of the Gangetic plain, are both of this nature. From such soils there is always a large amount of evaporation, even when their surface seems quite dry, and they neither heat nor cool rapidly. The sands of Sind and Rajputana on the other hand retain no moisture. What little rain they receive soon dries up, and in the hot weather they are perfectly dry for several feet below the surface. Under the summer sun, therefore, they heat with great rapidity and to a very high degree; so much so that it is said often to be possible to cook an egg by simply laying it on the sand in the sunshine at noonday. Winds almost always blow from colder to warmer regions, and are one of nature's chief ways of equalising temperature. How much India owes to seasonal changes of wind we shall see presently. We may notice here, however, a common diurnal change which greatly mitigates the summer heat along the coasts of the peninsula. During the early part of the day the air over the land is warmed by the sun's rays far more than that over the sea. In the course of the afternoon, therefore, a cool and refreshing breeze sets in from the sea which attains its greatest strength a little before sunset and is felt for many miles inland. The hotter the day and the drier the air, the sooner does the sea-breeze begin and the stronger floes it blow. A few hours after sunset it dies down, and then, if the night be clear the air over the land cools more rapidly than that over the sea, so that towards morning a land-wind is established blowing out to sea. The sea breezes which blow daily in the hottest weather greatly moderate the heat along the Coromandel coast.

That elevation reduces temperature is a very familiar experience in India, where cool refuges from the heat of the plains are found at a multitude of hill stations. The extent of this reduction is about 1° F. for every 270 feet of vertical ascent. The plateau of Mysore, being over 3,000 feet in height, is always from 10° to 12° F. cooler than the adjacent plains. The same cause greatly moderates the heat all over the Deccan. 85. Hills or mountain chains exercise in other was also a powerful influence on climate. The slope of the hills turned towards the sun is always much warmer than the slope that looks away from it. Mountain chains also often intercept winds, increasing, or sometimes, though more seldom, reducing, the temperature of the protected plains as a result. The Himalayas protect the Gangetic plain from the keen and icy north winds that blow in winter across Tibet. On the other hand, during the south-west Monsoon, the Western Ghats keep both wind and rain from the plains immediately to their east, where both would be more than welcome. In the light of the foregoing paragraphs we may now illustrate the prevailing temperatures of India by a series of isothermal charts. In the following charts the recorded temperatures are reduced to sea-level, i.e.. they are increased by 1° F. for every 270 feet of elevation. The approximate temperature of any place can therefore easily be found by dividing its elevation in feet by 270 and deducting that number of degrees from the temperature. It will be observed that in December, the coolest month of the year, the hottest part of India is a small tract inland from Goa, almost half way between Cutch and Cape Comorin, where the mean temperature is over 80° F. Next come the southern and western parts of the peninsula, then the eastern as far north as a line stretching from north of Bombay to Vizagapatam. From this point the isotherms stretch irregularly from east to west across India, colder temperatures prevailing as we pass further and further north. By March the sun has come north to the equator and the temperature has increased all over India, but the peninsula is still the hottest part. Along the coast the temperature is from 80° to 85° F., but a large interior tract is over 85° F., and within that is a smaller tract over 90° F. Taking India as a whole, May is the hottest month of the year, the sun being well on his way to the Tropic of Cancer and the south-west monsoon as yet hardly felt. The region of greatest heat, over 95° F., is in Central India, with a large tongue stretching into Rajputana.

The surrounding area, from 90° to 95° F., keeps clear of the coast except in the north-west, and stretches into Baluchistan. The coolest parts of India are a strip along the west coast and the whole region north and east of the Bay of Bengal. By August the full cooling effects of the monsoon have been felt, and it will be seen that the area of high temperature has moved away to the north-west, to a region untouched by the monsoon currents.

The Monsoons and Rainfall

The Monsoons are the seasonal winds that prevail in India and blow alternately from the south-west and north-east, bringing the abundant rains upon which the fertility and wealth of the country depend. The southwest monsoon, which in ordinary years reaches the southwest coast in May and blows in full force in the northern Provinces by the end of June, is by far the most important. It is emphatically the monsoon. It brings to five-sixths of peninsular and continental India and the southern slopes of their mountain wall their main supply of water. The north-east monsoon, though of less moment to the country as a whole, is of great importance to the south-east of the peninsula and the north of Cealon. These parts receive but little rain from the south-west, and the northeast monsoon makes up the deficiency.

The causes of the monsoons are not difficult to understand. We have seen that winds always blow from regions of higher to regions of lower pressure, and that regions of high temperature are also regions of low pressure. Winds therefore blow on the surface of the earth from colder to warmer regions. If the surface of the earth were all water, winds would blow in both hemispheres towards the equatorial belt, which would constitute a permanent zone of low pressure. By the rotation of the earth such winds would be diverted towards the west, and therefore north-east winds would prevail in the northern hemisphere and south-east winds in the southern hemisphere, and in each case these winds would be strongest when the sun was at the other side of the equator. As the southern hemisphere is mainly water, south-east winds do actually prevail over the greater part of its surface. But the northern hemisphere has more land than water, and therefore, owing to the different degrees in which land and water are heated by the sun's rays, the areas of lowest pressure are sometimes far removed from the equator, and the direction of the prevailing winds are changed. The winds in the northern hemisphere are therefore not characterized by the comparative uniformity that prevails in the south, and sometimes—as in the case of the south-west monsoon in India—the primal conditions are completely reversed.

By March 22nd the sun has passed the equator on his way north, and the whole of India, then everywhere fairly dry, is rapidly increasing in temperature. During that month the average pressure falls over the entire country. By the end of April an area of deeper depression has been formed over the United Provinces and Central India, and already over a considerable part of the peninsula light south-west winds have begun. These do not come from the sea, however, and so bring no water with them. By the middle of May the depression has largely increased in extent and its centre has become deeper. Its influence is consequently felt over a wider area, and as far south as the equator the winds are now mainly south-west. They strike the west coasts of India after travelling over warm oceans for many hundreds of miles, and come laden with moisture.

By the end of June the depression has increased enormously in extent, and has still further deepened over northern India. The monsoon is now fully developed, and south-west winds prevail for 30° north of the equator and from the coast of Africa to the Philppines. In July the depression is still larger and its centre of greatest depth is over Baluchistan. During August and September the depression gradually diminishes and moves to the south-east. The charts on page 55 illustrate the formation of the low pressure system and indicate the prevailing winds. By September 22nd the sun has passed south of the equator, and the vast dry highlands of Asia are cooling with extraordinary rapidity. The conditions of Central Asia as to pressure are therefore soon totally reversed. By the middle of October a large system of moderately high pressure has been formed, extending from the Caspian Sea to China. That remains of the old depression, greatly reduced, is now over the Bay of Bengal. North-east winds have begun to blow in the north of the Bay, though south-west winds still prevail in the south and east. Now winds blowing from the north-east come from colder and comparatively dry latitudes. They bring little moisture with them, and though they would take up a considerable quantity from the Bay of Bengal, they could give little rain to the east coast of India, upon which they blow. For they are travelling from colder to warmer latitudes, and, growing warmer as they go, are ever increasing their water-bearing power.

But south-west winds are still blowing in the south and east of the Bay of Bengal, and these coming from the warmer southern oceans are laden with moisture. Meeting the north-east winds this southerly current is bent round in the north of the Bay, and, chilled by the colder current which turns it back, it sheds its moisture in a fertilising flood on the east coast of the peninsula and far inland. It is for this cause that the north-east monsoon is sometimes called the "retreating monsoon," for its rain-giving power depends upon the south-west winds which it meets and overcomes. The rains of the north-east monsoon are, however, soon over, for by the end of November, the high pressure area having increased in intensity, north-east winds prevail over the whole Bay and the warm water-bearing currents are turned back long before they reach the latitude of India. The charts on the preceding page show the formation of the high pressure system over Central Asia, and the prevailing winds for October and December.

Though we speak of the "South-West Monsoon" and the "North-East Monsoon," thereby indicating the general direction of the wind, it must not be supposed that the winds actually maintain these directions in every part of India. They do not. They are diverted from their original direction to some extent by local variations of pressure, and to a much greater extent by the configuration of the country. The decline in barometric pressure from west to east during the south-west monsoon bends the south-west winds round towards the east, and they strike the coasts of Bombay as westerly winds. The opposite is the case in the north of Burma where the rapidly decreasing pressure to the north-west bends the winds round towards the north and then to the north-west. This influence is greatly intensified by the northern mountain wall, across which they cannot pass. The south-west monsoon current in the Gangetic plain blows, therefore, from the south-east, travelling from Bengal up the plain to the Punjab and then bending north again. For similar reasons the north-east monsoon in the Gangetic plain is a north-west wind, travelling from the Punjab to Bengal, where it bends round to the south. Blowing from the north-east in the northern part of the Bay, it' skirts the area of depression, bending gradually towards the south.

We must now follow the course of these winds and mark the conditions which determine the deposit of their life-giving store of water. The south-west winds, striking the west coast, are at once checked by the Western Ghats. To pass these the current is forced upward to colder altitudes, and at once a large part of its vapour is condensed, and heavy and continuous rains are the result from beyond the summit of the mountains to the sea. The rainfall is great all along the western coast during the whole of the south-west monsoon, but more particularly in the south, where it commences earliest. East of the Ghats the rainfall rapidly declines, for the current descends to warmer regions. In the month of July, when rain is falling heavily all along the Western Coast the winds reach Madras as hot and scorching blasts.

Towards the north the Western Ghats decline in height, and thus present less hindrance to the monsoon current. North of the Ghats are the long valleys of the Tapti and the Narbada. Up these the current sweeps, depositing a portion of its water as it goes, but reserving the bulk of it for the broken highlands of the Central Provinces and Chota Nagpur, where during the month of July the rainfall is very abundant. Further north again Kathiawar and Cutch get comparatively little, and Sind and the Indus valley very much less. The land is hot, and there are no mountains to force the current upwards, and so the winds sweep past, carrying their moisture with them till it is condensed on the slopes of the western Himalayas.

On the eastern side of India the south-west monsoon first gives a heavy rainfall to the south and west of Ceylon, but leaves the northern and eastern part of the island almost untouched. Sweeping up the Bay it strikes the coasts of Burma. Tenasserim receives a very copious watering, as also does the Arakan coast west of the Yoma. Fairly heavy rains, indeed, prevail for a couple of months over almost the whole of Burma, except a central area a couple of hundred miles north and south of Mandalay, which is comparatively dry. It is a warm and sheltered region consisting of plains and low hills, and the monsoon current, having deposited much of its water in the south and west, passes over it without shedding much more of its moisture till it meets the colder mountains of the north.

North of the Bay of Bengal, and in the valley of Assam, the rainfall is very heavy. Cherrapunji, on the southern edge of the Khasi Hills, has a fall far in excess of that of any other place in the world, receiving as a rule over 600 inches of rain in the year. This phenomenal fall, which extends only over a very small area, is due to a combination of causes. The clouds have swept up over many hundreds of miles of the warm bay and the Ganges delta, and the air is completely saturated with moisture. The slopes of the hills are sharp, and the current is swiftly diverted upwards. But this alone would not account for such a fall. Just at that point the current that is diverted toward the west by the frontier hills of Burma, meets that coming up direct from the bay, and it is probable that an upward swirl is caused that carries vast volumes of saturated air to cooler heights, ridding it thus of almost all its moisture. From the Khasis one branch of the southwestern current is diverted in a north-westerly direction along the southern slopes of the Himalayas and the great plain, and the other, turning to the east, passes up the valley of the Brahmaputra. Both these branches yield abundant rain, and the Provinces which they traverse are among the best watered tracts in India.

During the north-east monsoon the rainfall is chiefly on the east coast, extending inland in the south right across the peninsula to the Western Ghats. These are the districts to which the south-west monsoon gives but little rain, blowing over them for the most part as a comparatively dry and warm wind. The north-east monsoon makes up for this deficiency, and in the place of the summer rains which prevail elsewhere gives them a full supply in the autumn.

In April the south-west monsoon currents are slowly forming, and by the end of the month Ceylon and Travancore have received the beginning of their rains. In the diagram for this month, as well as in those for July and October, notice should be taken of the comparatively dry area in central Burma. In July the south-west monsoon is in full force, and it will be seen how greatly the Western Ghats keep the rain from the districts to the east of them.

In October the north-east (or " retreating ") monsoon is beginning. From the direct south-west currents rain is still falling on the south-west coasts of India and Burma, and the retreating current has begun to shed its store along the coast of Madras.

Taking the rainfall of India as a whole, we find that in most parts it is fairly abundant. The northeastern portion of the great Indo-Gangetic plain, the whole of Bengal and Assam, the greater part of Burma, and the West Coast Districts from the Gulf of Cambay to Cape Comorin, have a very copious supply. Next to them come the eastern Districts of the Deccan Plateau. These have as a rule a heavier fall than the western Districts, where the rains are much more uncertain and variable. In the north-west of India the fall is always scanty and insufficient, and much of the land is consequently little better than desert. East of the Indus a long narrow strip is almost rainless.

Though the climate and rainfall of India are thus subject to well ascertained laws, and are on the whole exceedingly regular, they are nevertheless liable, particularly at the change of monsoons, to local disturbances of peculiar violence. Thunder storms are very common in many parts, and, though quickly over, they often deluge large tracts of country over which they pass. They are as a rule very welcome. They bring relief from the oppressive heat, and their rains refresh the whole face of nature. In the Bay of Bengal Cyclones of great severity are also frequently generated, especially at the change of monsoons.

They seem, as a rule, to take their rise in the neighbourhood of the Andaman Islands, and travel at first in a westerly direction gradually changing their course to north and then north-east.

Cyclone storms are approximately circular or elliptical, and their centre is an area of deep depression. The wind blows round the centre, but bends ever inwards and upwards. Cyclones commonly strike the coast to the north of Madras, and, passing northwards, the fury of their winds often does great damage along the coast. They are usually accompanied by exceedingly heavy rains. Owing in part to the rapid decrease in barometric pressure towards the centre (which alone would account for a rise of 3 feet in the level of the water), and in part to the inward and upward motion of the wind, vast storm waves are sometimes formed, and when these strike a confined bay they sweep over the land for many miles, causing great destruction of life and property. About thirty years ago the town of Masulipatam was almost destroyed by such a wave, and a few years later a similar one swept over the Sandarbans and up the estuaries of the Ganges, doing immense damage and destroying over 100,000 lives.

NATURAL PRODUCTS

As is necessary preliminary to any useful study of the natural products of India, and especially of the cultivated vegetable products, we must first consider the nature of the soil and the supply of water for agricultural purposes. We have glanced at both these subjects in other connections, but they claim now a more detailed and careful examination if we would understand the natural distribution of vegetable products, and the reasons why different crops, or different methods of cultivation, prevail in different districts.

THE SOIL

The soil is the weathered product of the rocks of which the earth's crust is composed. Whenever rocks, however hard, are exposed to the influence of air and water, heat and cold, they slowly crumble. This process, which is partly chemical and partly mechanical, is termed "weathering." It is accelerated by the presence of vegetation, for not only do plants promote chemical action, but their roots, penetrating into the minutest crevices, soon split the hardest rocks. The soil is the product of long ages of such weathering, increased and enriched by the decay of plants and animals. Whatever other substances may have mingled with it and modified it, it is plain that the character of the soil must everywhere depend primarily upon the character of the rock from which it is derived. In a great alluvial plain like that of the Ganges the soil is a mixture of the crumblings of many kinds of rock which the rivers have brought from great distances, and which have been ground down by attrition to a fine impalpable mud. But in othercases there may be little mixture, and especially is this the case where particular geological formations cover large areas, when the difference between different soils is almost as marked as the difference between the rocks from which they are derived. Nowhere is this more clearly seen than in the Deccan, where the soil that prevails in the north-west offers a striking contrast to that of the cast and south.

We have seen that the Aravallis and the Eastern Ghats are the most ancient hills in India. The geological formation of which they are in the main composed extends right across the peninsula south of a line drawn from Goa to Masulipatam, and across the eastern half of that part of the peninsula which lies north of that line. Over the greater part of this area, and again in the west from the northern slurs of the Aravallis to the falls of kutch and Cambay, Jiard crystalline archican rocks prevail. Such rocks weather very slowly, and although the process of disintegration has been going on for countless ages, the soil is of no great depth excess where it has been washed down from the hill sides into the valleys. The soil from such rock is porous and light, and the rain sinks into it readily. But as the rock itself is comparatively near the surface and is impervious to water, the moisture that the soil receives drains away with rapidity, or is collected in deep hollows of the rock.

Between the Archaean region of the Aravallis and that of the south-west there is a vast area of basaltic rock, known as Deccan Trap, which is volcanic in its origin. Through long geological ages that part of India was the scene of vast and recurrent volcanic disturbances, the outflow of which ultimately covered an area of upwards of 200,000 square miles. It forms the entire north-western part of the Deccan and two-thirds of Kathiawar, and in successive layers extends to a very great depth, in some places exceeding 6,000 feet. Deccan Trap weathers comparatively quickly, and the resulting soil is dark in colour and very fertile, and is known as the black cotton soil. Unlike the soil from the crystalline rocks of the east it retains its water, so much so that it is often described as " water-holding soil." And since, owing to the rapid disintegration of the trap, the soil is generally fairly deep, there is usually a good supply of water a little way below the surface which can easily be tapped by wells.

The water-holding power of any soil is of the greatest possible value to the agriculturist. Where the soil has but little such power the effects of drought are very speedily felt, the fields are burnt up and the wells run dry. A water-holding soil, however, may be caked on the surface, but still vegetation flourishes, for as the surface dries the water rises from below. With such a soil it is only after many rainless months that the effects of drought begin to be seen.

The cotton soil possesses this power to a remarkable degree. To a less degree it is also characteristic of most of the alluvial soil of India. This is particularly the case in the valley and delta of the Ganges, where the soil is a mixture of clay and mud, and to a less extent also in the northern part of the great plain between the Jumna and the Indus, where the soil is a light, but not sandy, loam.

Passing south-west, however, down the Indus valley, the soil becomes more and more mixed with sand, till in Sind it is little else. The scanty rains which that region receives are quickly absorbed by the light and porous earth; but they evaporate almost as quickly, and the ground is soon perfectly dry for many feet below the surface.

IRRIGATION

By irrigation we commonly mean the watering of the land for the purposes of agriculture by water brought from a distance by means of canals from rivers, or from storage tanks. In many parts of India a sufficient supply of water can be obtained in ordinary years from wells alone. This is the case wherever water-holding soil prevails. Where wells can be depended on there is less need for other sources of supply, though even there a perennial supply of river water may be both cheaper and better for the land. Well-irrigation exists all over India, sometimes alone, and sometimes side by side with other systems, and, taking India as a whole, probably a greater area is watered by wells than by all other systems of irrigation put together. But well-irrigation is chiefly a private work, and land so watered is seldom, in the technical sense, irrigated land. It is probable that irrigation from wells may in some places be taken up by Government, experiments with power-pumps having been made in several taluks in the south. But so far little has been done.

The rivers and plains of North India are peculiarly suitable for the development of great systems of canal irrigation. The rivers being snow-fed afford even in their upper courses and in the digest seasons a fairly abundant supply of water. This can be drawn off into canals at the highest part of the almost level plain, and the gentle slope of the plain then gives the fall necessary for steady flow. 144. As an illustration of this system, and as showing the magnitude of the works that have sometimes to be undertaken, we may take the Upper Ganges Canal which has been in operation for upwards of half a century. The Canal head is near Hardwar, a few miles below the junction of the Bhagirathi and Alaknanda. At this point the Ganges is a fair-sized river, having a flow of about 7,000 cubic feet per second in the driest season, and much more in the rains, or when the snow has begun to melt. The vast head-works of the Canal are of solid masonry and are so arranged as to draw off about 6,500 cubic feet of water per second. This great volume of water is taken in a south-westerly direction across the course of other mountain streams which it does not disturb. It passes under one by means of a tunnel, and is carried over another by an aqueduct two miles in length. Then it bends to the south, and by means of main channels over 450 miles in length, and smaller distributaries with a total length of almost 4,500 miles, waters an area of 1,500 square miles of land between the Ganges and the Jumna. A little lower down, when the Ganges has again become a river of considerable volume, another canal takes off an almost equal quantity of water.

Similar canals are taken from almost all the main tributaries of the Ganges and Indus, and some are of even greater magnitude. The Sirhind Canal from the Sutlej waters more than 1,200 sq. miles of the Panjab, as well as large tracts in the Native States of Patiala, Nabha and Jind. Its main channel exceeds 500 miles in length, and has over ten times that length of distributaries. The Lower Chenab Canal has a main channel of 427 miles, and waters the large area of 3,040 square miles. The Jumna, the Ravi, the Jhelum and the Gandak, provide water for other canal systems which give a perennial supply to many millions of acres. In the Punjab alone 250 square miles of laud are thus watered. 146. In the canals already mentioned the supply of water is constant, the head-works being so constructed as to draw off a sufficient volume even when the river is at its lowest. In the lower course of the Indus a different system is adopted, the canals being tilled only during the lime with the river is in flood. Such canals are distinguished as Inundation Canals, the Indus, owing to the high level of its bed, offers special facilities for this system, which, though not affording so perfect a protection, has the advantage of cheapness. The solid masonry canal heads necessary for the perennial canals give place to simpler earthworks, and a much smaller capital expenditure is required. While the Upper Ganges Canal has cost over;£2, 000,000, and. the Sirhind and Lower Chenab Canals each not very much less, the Indus Inundation Canals, with over 650 miles of main channels, have cost less than;100,000. Sind is almost entirely dependent upon irrigation of this kind.

In the peninsula canal irrigation from the rivers is much more restricted than in the northern plains, since the rivers, not being snow-fed, do not offer a continuous supply of any great volume until they are comparatively near the sea. Small canals, however, draw their supplies from the upper reaches of the Godavari and Kistna; and the Son, shortly before its confluence with the Ganges, supplies canals which irrigate a considerable area in Bihar. But it is in the deltas of the rivers that the great irrigation works of the peninsula are found. The waters of the Mahanadi, the Godavari, the Kistna and the Cauvery are all thus utilised, as well as those of some of the smaller rivers. An anicat, or masonry dam, is thrown across the river near the apex of its delta, which prevents the water draining away too rapidly to the sea. The level of the water above the anicat is thus considerably raised, and a slight fall secured that enables it to be easily distributed throughout the delta and to some extent further inland. The irrigation systems that water the deltas of the four great rivers that discharge into the Bay of Bengal have over 2,000 miles of main canals and 5,300 miles of distributaries, and give an unfailing supply of water to over 4,000 square miles of exceedingly fertile land.

Irrigation in the interior of the peninsula is, however, chiefly from tanks. We have seen that the rivers are in flood for only a short time, and that over the whole of the eastern and southern half of the peninsula the ground is such that the water speedily drains away. The problem to be solved is, therefore, how to hinder the water that Hoods the land during the rainy season from running to waste. This is solved by storing it in " tanks " where it is available for future use. These tanks are of all sizes, from mere ponds to lakes five or six miles in length. They are usually constructed by throwing a dam or bund of masonry or earthwork across a narrow valley through which a stream passes, thus confining the natural drainage. Or sometimes these tanks, or lakes, are constructed at some distance from a river whose waters are artificially turned into them. The water is then distributed over the surrounding country by a network of channels.

In the Madras Presidency alone there are 60,000 such tanks of all sizes. Many of them are ancient works, but most of the greatest have been constructed in recent years. One notable illustration may be mentioned—the Periyar Project, as it is called. The Periyar is a river on the western side of the Western Ghats, whose waters used to be lost in the Arabian Sea. They are now diverted and carried through a tunnel under the Ghats to the eastern side of the hills where water was greatly needed. They supply a vast artificial lake capable of watering more than 300 square miles of land.

Indian irrigation works surpass in extent and utility anything else of the kind in the whole world. The total capital outlay upon them up to the end of 1906 exceeded £3 1,000,000. The payment received from the cultivators for the water supplied meets all the working expenses and returns a fair interest on this large sum. The extent to which such works increase the wealth of the country, and especially of the ryots, is best seen from the fact that in seasons of only slight scarcity the value of the crops raised on irrigated land, and which, but for irrigation, could not have been raised at all, exceeds the whole capital outlay on the works themselves.

FORESTS

The forests of India constitute a valuable part of the natural resources of the country, because of the limber they provide, and they arc of further importance because of their influence on climate and rainfall. Forests protect the hill-sides, the roots of the trees binding the soil and hindering it being washed away. They also check evaporation, and so preserve the moisture of the soil. And, what is of still greater moment, wherever a large extent of forest occurs its comparatively cool area is frequently sufficient to attract the clouds and determine a downpour of rain, which, when once started, spreads far beyond the actual forest area. Forests are of importance for another reason. They encourage and protect the undergrowth of grass and small shrubs, which constitute an invaluable grazing ground when vegetation on the more exposed land is burnt up.

If tradition is to be believed the forests of India covered at one time the face of the whole land. But for centuries they have been exposed to indiscriminate destruction. No steps were taken for their preservation till 1846 when conservancy operations were begun in Bombay. Ten years later this example was followed in Madras. In 1861 the Forest Department of the Government was created, and the work thus begun was completed by the Forest Act of 1878, which gave to India a complete, scientific, and efficient system of forest administration. In all the Provinces there are now large areas of " reserved " forests, which are entirely under the control of the Department, as well as other, and in some provinces considerably larger, areas which are demarcated and efficiently protected. The objects which the Department has in view are (1) the protection of such forests as now exist from damage through unscientific felling of timber, or from fires; (2) the extension of the forests by planting suitable and useful trees over areas reserved for the purpose; (3) the production of as much good timber and firewood as the forest can yield without injury; and (4) the provision of grazing areas which can be relied on in times of drought. The total demarcated forest in British India is over a quarter of a million square miles, or about twice the area of the British Isles.

Numerous valuable timber trees are native to the forests of India. Of these by far the most important is teak. The tree is found chiefly in the forests of Burma and the Western Ghats, in both of which regions much attention has been given to its cultivation. About 150,000 tons of teak, are exported from Burma yearly. It is floated in vast rafts down the rivers, particularly the Salwin. Teak is a hard and durable wood, and until it is very old is not attacked by white ants. It is, therefore, specially useful for building purposes and for furniture. The sal is found in great abundance in the forests of the Himalayas and the Central Provinces. Its wood is hard and heavy, and is used for building purposes and railway sleepers. The sissoo is also characteristic of the Himalayan forests. Its wood is of a rich dark brown colour, hard and capable of a fine polish, and is used for furniture. The blackwood is found chiefly on the Western Ghats. Its heart-wood is a deep reddish-black, hard and firm.

The sandal grows in the drier parts of the peninsula, especially in Mysore. It is cultivated by Government in the Central Provinces. The heart-wood, has a lasting fragrance, and is much used for carving, and as a perfume. It is also employed in the manufacture of incense, and is exported for this purpose. The khair and the toon, both of which are common in the N.W. Himalayas and Burma, yield red woods used for furniture. The heart- wood of the former is dark in colour and very hard and durable, and is valuable for building. The deodar, a kind of cedar, along with various species of pines, are the chief trees in the higher forests of the Western Himalayas. The ebony tree grows on the Western Ghats, and in Burma the ironwood tree is next in importance to the teak.

The forests yield other important products besides timber. From the heart-wood of the khair cutch, or catechu, an astringent gum-resin used both in tanning and as a medicine, is obtained. Myrobalans, the dried fruit of several species of terminalia, are also a valuable tanning material. They are exported in considerable quantities. Caoutchouc, or India rubber, is obtained from the milk which exudes from incisions in the stem of the rubber tree, a species of fig, which is found in the eastern provinces of the Empire. Great attention has been paid of late to the cultivation of this tree, and Government plantations have been started in Bengal, Assam, and Burma. Two other species of the same genus, the banyan and the peepul are common in most parts of India. They, are large and handsome trees, but are not of any great economic value. The peepul is a sacred tree among the Hindus. The bamboo, a giant grass, is common in almost all parts of India where water is plentiful, up to an elevation of 3,000 feet. It is an important forest product, the reserved forests yielding in a single year nearly 100 million canes.

Many trees not indigenous to India have been introduced of recent years. The most notable of these are various species of eucalyptus which have been introduced from Australia and of which there are now extensive plantattons on the Himalayas, Nilgiris, ami Palnis, where they flourish greatly. The leaves of these trees contain an aromatic resin, and their cultivation is said to counteract malaria. The tree grows to an immense height, but develops so quickly that its wood is of little use except for fuel. The pain tree is also spreading in the hotter plains. It grows with great rapidity, spreading its branches over a wide area, and giving a thick and welcome shade. The casuapina is another quick-growing tree, which is cultivated in many places along the coasts of the peninsula. Large Government casuarina plantations have been established on the east coast. The straight poles of the trees are used for scaffolding, but the chief value of the casuarina is as a source of excellent firewood.

FOOD GRAINS

Rice

Of all the food grains of India rice is the most important. It is the staple food of more than a fourth of the people, and a common article of diet of at least as many more. Rice requires for its cultivation an abundant supply of water and warmth, and is therefore chiefly grown in those districts on the plains which have a copious rainfall or are well irrigated. The rich wet plains of Bengal form one of the largest and most productive rice-fields in great rice- producing province, but its total output is less than one-fourth that of Bengal. In both these provinces rice is the main food of the people, but while the vast population of Bengal consumes almost the whole of the rice grown in the province, three-fifths of the crop of Lower Burma is available for export. Rice is also extensively grown in the deltas of the peninsular rivers, and wherever the conditions of the country are favourable. Wheat is cultivated largely in North and Central India, and is a cereal of increasing importance, both as a staple food of the people and as an article of export. It requires for its cultivation much less water than rice. When young it can stand keen frosts, but after the ear is formed it needs a dry air and bright sunshine to bring it to perfection. The plains of Northern and Central India, especially towards the west, are thus well adapted for its growth. Indian wheat is hard and of excellent quality, and is growing in favour in Europe. Its cultivation is, therefore, spreading, and its export rapidly increasing. In the year 1907 the extent of land devoted to wheat culture in British India was 25 million acres, against 73 million acres devoted to rice. The total rice crop was about 21 million tons, and the wheat crop exceeded 8 million tons.

Millets

The chief millets grown in India are cholum (or jowar), cumbu (or bajra), and ragi. They require much less water than rice and take the place of that grain in most of the drier provinces of India. In all but the great rice-producing districts of the peninsula millets form the staple food of the poorer people, rice being a luxury of the wealthy. In Bombay, Sind, and Berar more than half of the total area devoted to food grains is given up to millets, and more than one-third in Madras and in at least half the Districts of the Punjab, Agra, and Upper Burma. Taking British India alone the millets are not so important a food crop as rice, but if the Native States be added, the majority of which are less abundantly watered than the British Provinces, they are more important, and form the staff of life to a larger number of people.

Pulses

Various pulses, the chief of which are gram and dal, are widely grown in the United Provinces and the Punjab, and less extensively in other provinces. They are valuable as foods, being more nutritious than either rice or millets, because of the larger proportion of nitrogenous matter they contain. They are eaten in combination with less nutritious grains almost all over India.

Barley is grown in the United Provinces, and to a less extent in the Punjab and Bengal, It is less nutritious than the millets, but is cheaper, and is, therefore, largely used by the poorer classes. Barley is also extensively used for brewing.

Maize, or Indian corn, is also cultivated in these provinces, and to a less extent in many other parts of India. It is nowhere, however, a crop of first importance.

OTHER VEGETABLE PRODUCTS

Palms

Various species of Palms are common throughout the plains of India, and on the hill sides, though being an essentially tropical order they do not flourish at any great height. By far the most useful of these is the cocoanut. It loves a well-watered, sandy soil, and flourishes all round the coasts of the peninsula, and particularly around the backwaters of Cochin and Travancore. It is also largely cultivated in the interior. The chief products of the cocoanut are copra and coir. Copra is the dried kernel of the nut from which coconut oil is expressed. Coir is the fibrous husk of the nut, which is woven into coarse matting and rope. There is a considerable export of both copra and coir from the ports of the West Coast. The palmyra palm which is common all over the peninsula, but cannot stand the cold nights of the north-west, is chiefly of value as a source of " toddy," which is the sap drawn from the flower stalk and slightly fermented.

The areca palm, cultivated chiefly in Bengal but found almost wherever the cocoanut grows, yields a nut which is chewed all over India along with the leaf of the betel, or pepper vine. A species of date palm, known as the bastard date, is found all over India, particularly in Bengal, and is one of the chief sources of jaggery, or native sugar. The true date, the fruit of which is of great value in Arabia and North Africa, needs a dry, hot climate; and brings its fruit to perfection only in Sind. The wood of all the palms is used for temporary buildings, and the leaves make an excellent and durable thatch.

FRUITS AND VEGETABLES

Many kinds of fruit are grown in India, the most universal and useful of which is the plantain or banana. The mango, one of the most luscious fruits in the world, is abundant, and much attention has in some places been given to its culture. Oranges of excellent quality, limes and figs are cultivated in many parts. The jack fruit, the papaw, custard apples, guavas, pomegranates, melons, and pineapples are also common. On the hills many kinds of European fruit have been introduced — apples, pears, plums, strawberries, etc.—and in some places with fair success.

The same is also true of European vegetables. Peas, beans, cabbage, cauliflower, etc., are grown on the hills and on the Deccan plateaux, and potatoes have taken kindly to the country and are now grown even on the plains. The vegetables native

Spices and Condiments. Of these India has no great variety. Chillies and turmeric arc grown in most parts and arc universally used. Coriander, aniseed, ginger, and cummin are also cultivated. Several species of pepper are grown along the Malabar coast strip, and in Travancore cardamoms are a valuable Government monopoly.

Opium is a powerful narcotic drug obtained from a species of poppy. Shortly after the plant has flowered incisions are made in the green capsule. The juice which exudes solidifies on the outside of the capsule, from which it is daily collected. Cleaned and further dried this exudation constitutes the crude opium of commerce. Opium is an exceedingly valuable medicine. It is also widely used as an article of vicious indulgence. For this purpose it is commonly smoked, or small quantities of it are swallowed, or an infusion is made and drunk. In whatever form it is used it acts first as a stimulant and then as a powerful narcotic and soporific. It has long been widely used as a luxury by many classes in India, notably the Sikhs and Rajputs; and, in parts of Bengal, Assam, and Burma, it is relied on as an antidote to malaria.

Following the example of the Mughal Emperors the British Rulers of India early made opium a Government monopoly. In British India it is produced chiefly in Bihar and the United Provinces, where the cultivators grow it under official inspection, the Government making advances on the crop, the whole produce of which is handed over to their agents.

The central Government Opium Depots are at Patna and Ghazipur. There the opium is packed in chests and forwarded to Calcutta, where it is sold by auction for export. Opium is also grown in the Native States of Central India and Rajputana. From some of these States large quantities of the drug are despatched to Bombay for export. This is known as Malwa Opium, and is subject to a very heavy tax as it passes through British territory. The area devoted to the culture of opium in the Ganges valley is about 600,000 acres, and the value of the crop is usually nearly 5 millions sterling. About 93 per cent, of the entire produce is exported, chiefly to China.

Tobacco is not a native of India, but was introduced by the Portuguese. It is now grown and used in every province, but most extensively in Bengal, Madras, and Burma. The area under tobacco in British India in 1907 was over a million acres, more than half of which was in Bengal. Much of the Bengal tobacco goes to Burma, where smoking is a universal habit. A small quantity of manufactured tobacco is exported to Europe.

ETHNOLOGY

The ethnology of India is still in its infancy, and only its bare outlines can be regarded as in any degree settled. In some respects it is less settled now than it appeared to be a few years ago. By modem ethnologists more reliance is placed on the study of physical types than on that of language, and many conclusions that were believed, on linguistic evidence, to be firmly established are now widely rejected. Twenty years ago language was regarded as yielding by far the most valuable testimony as to the origins of races and tribes and their pre-historic movements. At the present time a minute study of physical type is more relied on and in this direction a large amount of most valuable material has been gathered by the Ethnographic Survey of India, commenced in Bengal twenty years ago and now proceeding in most of the Provinces. The results of this survey up to 190 1 were carefully worked out in the General Report on the Census of that year.

But before summarising these results, it will be well to glance briefly at the theories commonly accepted till within the last ten years. In the main these are supported by recent investigations, but in some important points they must now be greatly modified. The vast majority of the people of India have long been known to belong to two great stocks, the Aryan and the Dravidian. To what extent these two stocks have mingled, where the purest blood of each is now to be found, and over what extent of the country each now predominates —these are questions on which a variety of opinions have been held, and in which a study of type alone can give a decisive answer.

In addition to these great stocks, three others of subordinate importance to India have been distinguished. The Tibeto-

Burman, which has influenced the country on the east; the Scythian, of Mongolo-Tartar origin on the west; and the Kolarian, of doubtful origin, but closely related to the Dravidian, and represented in modern days by scattered tribes east and west of Central India. That the Aryans entered India from the northwest, pressing on in successive swarms probably through many centuries, has long been universally agreed. Language, custom, and tradition alike bear testimony to this. Wliether the Dravidians and Kolarians had similarly an external origin, and entered India in invading swarms long anterior to the inroads of the Aryans, or whether they should rather be regarded as the true aborigines of the country, has been a disputed point. As regards the Dravidians, the former conclusion has been the one most favoured. There exists in Baluchistan a " linguistic island " of Dravidian speech, the Brahmi language, cut off from the main body of Dravidian tongues by -nearly 1,000 miles. This has been taken as strong proof that the Dravidians, like the Aryans after them, came from the north-west. Sir W. W. Hunter accepted this, and further held that the Kolarians came into India from the east, and, stretching across the north of the peninsula, peopled the highlands from Orissa to the mouth of the Narbada before the influx of the Dravidians; and that the latter, when pressed by the Aryans in the north, broke through the Kolarian line and forced their way into the south of the peninsula. The Dravidians had thus driven " an ethnic wedge " down the centre of the peninsula, dividing the Kolarians into two sections. He thus accounted for the separation of the Bhils, Kolas, etc., of the west from the Santhals, Kols, etc., of the east—all of which tribes he believed to be Kolarian.

To these four stocks a fifth was added almost in historic times. The Scythians appear to have come from Central Asia, and to have forced their way across the northwest frontier. Their influence is to be traced throughout the whole of west India, but Hunter believed that this race had permanently occupied the plains of the Indus, and that the noble race of Rajputs, and the Jats, the most important agricultural tribe of western Punjab, are their modern representatives.

Such were the conclusions drawn mainly from a study of language, with whatever additional light was to be had from a comparison of social systems, from tradition, and from a general agreement of type. modern ethnologists, however, hold that physical type is far more persistent and unchanging than language, and that, when we can decipher it, we shall find the ethnic history of every people more truly recorded in their physical characteristics shape of head, style of features, stature, hair, eyes, etc. than in their speech. They further hold that a minute study of physical type is likely to prove of peculiar value in India, where for centuries tribes and castes have lived apart, with but little intermixture of blood; for in such a case types may be expected to persist with unusual definiteness. It is exceedingly probable that this line of investigation may yield most valuable results in India within the next few years. In the Census Report Mr. H. H. Risley, C.I.E., summarises the results of the Survey as far as it had then proceeded, and on the basis of the data collected divides the people of India into the following seven distinct types :

I. The Turko-Iranian type, in Baluchistan and the Northwest Frontier Province. Stature above wean; complexion fair; eyes mostly dark, but occasionally grey; hair on face plentiful; head broad; nose moderately narrow, prominent, and very long.

II. The Indo-Aryan type, in Punjab, Rajputana and Kashmir. Stature mostly tall; complexion fair; eyes dark : hair on face plentiful; head long; nose narrow and prominent, but not specially long.

III. The Scytho-Dravidian type of Western India. Head broad; complexion fair; hatr on face rather scanty; stature medium; nose moderately fine and not conspicuously long.

IV. The Apyo-Dravidian type in the United Provinces, parts of Rajputana, in Bihar, and Ceylon. Head long or medium; complexion from light brown to black; nose from medium to broad; stature usually below the average.

V. The Mongolo-Dravidian type of lower Bengal and Orissa. Head broad; complexion dark; hair on face usually plentiful; stature medium : nose medium, with a tendency to broad.

VI. The Mongoloid type of the Himalayas, Assam, Nepal, and Burma. Head broad; complexion dark, with a yellowish tinge; hair on face scanty; stature below average; nose fine to broad; face characteristically flat; eyelids often oblique.

VII. The Dravidian type of Madras. Hyderabad, the Central Provinces, most of Central India and Chota Nagpur. Stature short, or below mean; complexion very dark, approaching black; hair plentiful, with an occasional tendency to curl; eyes dark; head long; nose very broad.

It will be seen that the Aryan element is much less prominent than has hitherto been thought, and on the other hand, that the Dravidian element is more widespread. The purest Aryan type is found under Type IV. Among the Rajputs, Khattris, and Jats, and the theory that would assign a Scythic origin to these races may be regarded as finally disposed of.

They have apparently maintained very largely the purity of their blood, while the Indo-Aryans of the United Provinces have commingled with the Dravidians. In explanation of this Mr. Risley surmises that they came into India by slow and peaceful migration, bringing their women with them, while the ancestors of their western neighbours, coming into India at a later date and by a more toilsome and hazardous route, brought few, if any, women with them, and took wives of the daughters of the land. With regard to the Dravidians, Mr. Risley's conclusion is that they are the earliest inhabitants of India of whom we have any knowledge. It must be remembered, however, that each type gradually shades off into the neighbouring one, and that the boundaries are therefore only approximate. The divisions, moreover, only indicate the general type of the bulk of the people. The proportions in which the racial elements combine in different classes of the comiminity arc also very various. Among the Scytho-Dravidians the Scythic element predominates in the Baratha Brahmans, and the Aryan element predominates among the Brahmans of the Aryo-Dravidians. There is also a certain admixture of the Arjan element almost everywhere among the higher castes, and traces Scythic and Mongolian blood are found among Dravidians of the south.

DISTRIBUTION OF THE POPULATION

India is essentially an agricultural country. According to the last Census nearly 200 millions of the people were engaged directly in agricultural or pastoral occupations. These people arc necessarily scattered over the land and not congregated together in towns. Many more are indirectly employed on the land, being engaged in ministering to the needs of the agriculturists proper. These also are scattered over the country. India has no great mineral resources, and even what she has are as yet but little worked. Nor are there any great manufactures to draw the people together in towns. Small manufactures there are in plenty, and some have their centres in particular localities. But in no sense can India, or any Province of India, be called a manufacturing country. The result of this is that the vast majority of the people live in hamlets or villages, and the towns and cities are comparatively few, and for the most part small. This is best seen by comparing.

India with a great manufacturing country like England. While in India, out of a population of 294 millions, only 270 millions—or considerably less than 10 per cent.—live in towns of 5,000 inhabitants and over, in England 77 per cent, of the total population live in such towns. In the one case over 90 per cent, of the people are scattered over the land, but in the other only 23 per cent. India is thus emphatically a land of villages.

In ancient days these villages were generally selfcontained and self-organised communities. They had but little communication with the outside world and needed little. Living upon the soil, the majority of their people were cultivators, but the simple handicrafts, etc.. necessary for the independent life of an agricultural community were represented in every village, and all occupations passed from father to son. In ordinary years, when nature was propitious, such villagers had few inducements to look beyond their own narrow borders. Roads were few; travelling difficult and dangerous; and, except when some religious festival drew them to some famous shrine, they spent their quiet and laborious days among their own people. In many parts of India this is still to a large extent the case, but in others it is rapidly passing away. Better roads, and other means of communication, the spread—even among the illiterate—of some knowledge of the world beyond, the increase of their means, and the opening of outside markets for their produce, are all enlarging the outlook of the people, breaking down their village exclusiveness, and bringing them into touch with a wider world. But these influences do not tend to substitute town life for village life, nor will they ever do so. So long as India remains, as she must remain, an agricultural country, the vast majority of her people must live upon the land they till.

Cinchona

The Cinchona tree is grown for its bark, which is the source of quinine, the most useful of all febrifuges. Cinchona was introduced into India in i860, prior to which time it was almost confined to South America. There are now large Government Cinchona Plantations on the Nilgiris and at Darjeeling, and numerous private ones. The tree is also grown on many coffee estates, being planted between the coffee bushes. At the factories on the Government plantations quinine and a mixed febrifuge are manufactured and are supplied to the public at a cheap rate. Quinine is also supplied to the public through the Post Offices, where it is sold in small pice packets. An abundant supply of a cheap and effective febrifuge is of the greatest importance where fever is so prevalent as in India.

Indigo is a dark blue dye extracted from the leaves of a small annual plant by maceration in water. The indigo crop used to be of great value, but the dye is being rapidly superseded by a chemical product which is much cheaper though not so good. Indigo culture is consequently rapidly declining. Between 1880 and 1890 the value of the indigo exported from India averaged over 2 J millions sterling. It has now fallen to about one-fifth of that amount. Indigo is chiefly grown in Bengal, Madras, the United lrovinces and the Punjab, four-ninths of the entire amount produced being grown in Bengal, and half the remainder in Madras.

Cotton is the soft fibre which enfolds the seeds of a small annual plant which has been cultivated in India for many centuries. It grows with great luxuriance on the rich black soil of the Deccan Trap, which has thus earned the name of " cotton soil." Cotton is one of the most important agricultural products of India, for it is not only one of the main articles of export, but, what is of still greater moment, is the raw material of a considerable and growing local manufacture. Great efforts have been, and are still being made to improve the quality of Indian cotton. At present it is not equal to the American variety, which has a longer fibre and is therefore more easily worked, and of greater strength. Careful cultivation, and the introduction of the American species, will probably overcome this difficulty and make Indian cotton equal to any in the world. Cotton cultivation is steadily increasing in India. At present, about 10 million acres are devoted to it, 14 millions of which arc in British India and 5 millions in the Native States. The value of raw cotton exported in 1907 was over /14, 600, 000. The greatest cotton-growing Provinces are Bombay (including the Native States), Berar, and the Nizam's Dominions, but it is extensively grown also in Madras, the United Provinces, the Punjab, and the Central Provinces.

Jute is the fibre of a quick-growing herbaceous mallow, which usually attains a height of eight to ten feet within four months of sowing, and often reaches twelve feet. It is then cut and the stalks are left for some weeks to soak in water when the fibre is easily extracted and cleaned. Jute is grown almost exclusively in the Gangetic delta, where over 3 million acres are devoted to it. Although it exhausts the soil upon which it grows to such an extent that it is usually allowed to lie fallow one year in four, it is nevertheless a most valuable crop. Not only does the fibre realise remunerative prices, but the plant can be grown on land exposed to such severe floods that any other form of cultivation would be extremely precarious. The value of the export of raw and manufactured jute in 1907 was over 25,000,000.

In modern days the growth of towns on particular sites is almost always due to one or more of three reasons. This last is an essentially modern cause of towns, and is due to the increase in the facilities of travel, and of the wealth and leisure of the people. It is naturally, therefore, most operative in Europe and America, where a multitude of towns trace their growth and prosperity to no other cause. But it has also been operative in India, and such hill stations as Simla, Darjeeling, Ootacamund, etc., owe their existence to it alone. Such towns are, however, of less importance than those whose origin is to be ascribed to the other causes named. Though they may grow to considerable size and wealth, and may be, as in India, summer seats of Government, 'they exercise no great influence on the development and destiny of a people.

But while in these modern days most towns are indebted to commerce or manufacture for their prosperity, very different reasons may have led originally to their foundation. In ancient times (i) political and (2) religious considerations had more to do with the founding of towns in India than had commerce. At various periods the country has been overrun by foreign invaders, and at others has been broken up into a multitude of States almost ceaselessly at war with one another. In such times, wherever a powerful chief settled, people flocked to him, partly for defence and partly for trade; and a town soon grew up, which became his capital and the seat of his government. The site of such a town was chosen not so much for commercial as for military reasons. It had, indeed, to be situated, if possible, in some place to which the supplies needed by his retainers and army could be easily brought; but it was still more needful that it should be in a good strategic position, well adapted both for defence and as a base for attack. A large number of Indian towns had this origin. Many others owed their origin to religion. The presence of a famous shrine, or the proximity of a sacred river, attracted annually multitudes of pilgrims, whose requirements in the way of accommodation and provisions afforded lucrative employment for a large resident population.

When once a town has been founded, no matter what its origin may have been, tends to build up for itself a trade and thus to maintain itself in prosperity, even though the circumstances in which it had its rise should quite pass away.

Religious change in India is so exceedingly slow, that towns which grew up at first for the convenience of pilgrims are often preserved in prosperity for many centuries by the conditions which gave them birth. We have illustrations of this in Benares and Puri. Puri exists solely for the sake of pilgrims, and though Benares has now an additional importance, due to other and modern causes, yet its shrines and pilgrims are still the chief sources of its fame and wealth.

Political changes in India have, however, been as rapid as religious changes have been slow, but British rule has now given peace to the whole country. Many towns, therefore, that had a strategical origin are now flourishing trade centres. In most cases they possessed from the first, as we have seen, a certain suitability of position for trading purposes. When they became centres of environment, and seats of a large population, roads were pushed out in every direction, and other means of communication opened up, till in course of time they became the recognised emporia for large districts. When their political importance declined, their trade still maintained them and became a more enduring cause of prosperity and wealth. Under British rule such towns have frequently become centres of civil or military administration, and have regained much of their old greatness.

BIBLIOGRAPHY

Agnew, John, David N. Livingstone, and Alisdair Rogers: *Human Geography: An Essential Anthology.* Oxford and Cambridge, Mass.: Blackwell, 1996.

Bernauer, Thomas: *Genes, Trade, and Regulation: The Seeds of Conflict in Food Biotechnology*, Princeton, Princeton University. 2003.

Clark, Michael J., Kenneth J. Gregory, and Angela M. Gurnell: *Horizons in Physical Geography.* Basingstoke and London: Macmillan Education, 1987.

Cosgrove, Denis: *Apollo's Eye: A Cartographic Genealogy of the Earth in the Western Imagination.* Baltimore and London: John Hopkins University Press, 2001.

Costanza, R.: *Ecological Economics*, Columbia Uni. Press., New York, 1991.

Diamond, Jared: *Guns, Germs, and Steel: The Fates of Human Societies.* New York: Norton, 1999.

Daubenmire, F.: *Plants and Environment*, New York, Wiley, 1947.

E. Dereniak and D. Crowe, *Optical Radiation Detectord*, John Wiley and Sons, 1984.

Edward A.: *Papers on Bacterial Genetics*, Boston, Brown and Company, 1960.

Entrikin, J. Nicholas, and Stanley D. Brunn: *Reflections on Richard Hartshorne's "The Nature of Geography."* Washington, D.C.: Association of American Geographers, 1989.

Haggett, Peter: *Geography: A Modern Synthesis.* 2nd ed. New York: Harper and Row, 1975.

Hartshorne, Richard: *Perspective on the Nature of Geography.* Chicago: Published by Rand McNally and Company as Monograph No. 1, 1959.

Harvey, David: *Spaces of the Capital: Towards a Critical Geography.* Edinburgh: Edinburgh University Press, 2001.

Katz, Cindi, and Janis Monk: *Full Circles: Geographies of Women over the Life Course.* London and New York: Routledge, 1993.

Livingston, David N., and Charles W. J. Withers: *Geography and Enlightenment.* Chicago: University of Chicago Press, 1999.

Quaino, Massimo: *Geography and Marxism.* Translated by Alan Braley. Totowa, N.J.: Barnes and Noble, 1982.

Thrower, Norman J. W.: *Maps and Civilization: Cartography in Culture and Society.* Chicago and London: University of Chicago Press, 1999.

Teich, M.: *Fundamentals of Photonics*, New York: Wiley, 1991.

Walden, Richard: *Genetic Transformation in Plants*, England, Open University Press, 1988.

Wood, E.J. Ferguson: *Marine Microbial Ecology*, London, Chapman and Hall, 1965.